Environmental Injustices,

Political Struggles

David E. Camacho, Editor

ENVIRONMENTAL INJUSTICES, POLITICAL STRUGGLES

Race, Class, and the Environment

Duke University Press Durham & London

1998

Printed in the United States of America on acid-free paper ∞
Typeset in Adobe Garamond with Frutiger display by Keystone Typesetting, Inc.
Library of Congress Cataloging-in-Publication Data appear on the
last printed page of this book.

DEDICATED TO DIEGO AND GUILLERMO

CONTENTS

David E. Camacho

INTRODUCTION

In 1987, the United Church of Christ Commission for Racial Justice published the most widely recognized study of race and the incidence of environmental hazards in the United States. Nine years after the first complaint of "environmental racism" was raised, African Americans, American Indians, Asian Americans, and Latinos are even more likely to find themselves neighbors of commercial hazardous waste facilities. Moreover, local government officials consistently allow for the location of hazardous and toxic waste facilities in low-income neighborhoods and communities. Consequently, low-income groups as well as people of color suffer disproportionately from the regressive impacts of environmental policy. Further, the average fine for violations of federal hazardous waste and environmental statutes in these communities has been found to be significantly lower than those imposed for violations in largely white neighborhoods. These policy practices are due, in large part, to the persistent underrepresentation of people of color in the policymaking process and to the lack of public advocates who represent low-income neighborhoods and communities.

An unprecedented period in the history of environmentalism in the United States is upon us as low-income groups, members of the working class, and people of color steadily form an active and visible presence in local environmental activities. These groups are emerging as a movement for environmental justice and are providing the leadership for environmental activism in communities that have historically been neglected by environmental politics. Environmental concerns of the past—preservation and conservation—are being supplemented by struggles for participatory democracy. Much of the success of the environmental justice movement has come from the intersection of race, gender, and class groupings. This socially inclusive, multiracial coalition connects environmental issues with those of racial and gender inequality, lack of health care and social services, inadequate housing, poverty, and other economic barriers that have been the focus of the civil rights and social justice movements.

This book is especially designed with an eye toward the difficulties that accompany the topics of race, class, and the environment. *Environmental Injustices, Political Struggles* is a response to the lack of attention given to the political and social aspects of environmental problems. Technical, scientific issues receive little attention in this text. Rather, we aim to offer a fresh look at a set of issues dominated by "scientific" thinking. This book provides an alternative, and what we believe to be a more useful, perspective on environmental justice than the empiricist approach common to much policy analysis. We suggest that inquiries into social and political phenomena should be recognized as developing from the influence of a historical, subjective context and from societal presumptions. Our perspective asserts that the distinction between what can be known and measured (facts) and that which is not verifiable (values) is particularly limiting in thinking about and resolving complex social problems.

We invite political scientists, policy analysts, and students to ponder the value-laden character of public policy. Public policy is a political act, serving some actor's interests and promoting that actor's value judgments. Consider, for example, that policy analysis should address ethics and normative theory *and* the apparent normative basis of the status quo in the decisional process—that is, the values and interests represented in the existing regime and policy process. Consider, further, that sources of power must be critically examined. The study of environmental injustice must take into account the hierarchies of power that are inherent in a society. This point speaks to the importance of *context,* whereby any subject would be better understood if it were examined within the setting that both produces it as a subject and brings it to light as something to be studied. We attempt to place individuals within a "frame of reference" that provides guidelines for interpretation, explanation, prediction, and evaluation. The choice of a certain frame of reference makes some topics of inquiry more central and salient than others; it makes some kinds of policy instruments look markedly more attractive than others; it makes some social consequences more legitimate than others and, thus, affects the likelihood that public sector resources will be harnessed to their pursuit. In choosing a frame of reference, the individual picks not only an agenda for research but also an agenda for public policy. The individual is choosing among social values, and because values underlie decisions, the individual should recognize that the choice of a frame of reference is an ethical matter and not just a technical, scientific concern.

This last point is important for examining environmental justice. The issue of environmental justice is a new one, but one that is of interest to individuals from a wide range of backgrounds and professions, including political science, sociology, urban planning, law, and medicine. It deals with environmental problems (e.g., waste facility location, urban industrial pollution, and pesticides) that have a disproportionately adverse impact on lower socioeconomic classes and people of color. There are disagreements about the causes of this disparity. Some claim "environmental racism"; others maintain that policy decisions reflect market dynamics. These disagreements stem from different frames of reference. We ask our readers to bear in mind the following question: Is there a frame of reference that is suitable for addressing environmental justice? The articles in this anthology provide material leading to insightful discussions of environmental justice in the United States. Political, economic, social, and cultural factors are examined for their impact on the causes of environmental problems, especially as they affect people of color and low-income groups. As Stephen Sandweiss (this volume) states:

> Although a relative consensus exists concerning the definition of the problem of environmental injustice, there remains considerable disagreement regarding its causes and possible solutions. This disagreement seriously threatens the ability of the environmental justice movement to achieve the substantive policy changes it is demanding.

Our objective is to make readers *think* about environmental injustices and their possible remedies. Might, for example, the causes and consequences of environmental justice be located in hierarchies of political and economic power in the United States?

We begin *Environmental Injustices, Political Struggles* by posing the question, What is politics? in order to stress the political variable and, more important, to place the articles into a theoretical framework for inquiry. This framework is informed by the political process model, which is based on a particular conception of power in the United States consistent with the perspective advanced by an elite model of the political system. Like the elite model, the political process model rests on the assumption that wealth and power are concentrated in the hands of a few groups, thus depriving most people of any real influence over the major decisions that affect their lives. Accordingly, social movements—like the environmental justice movement—are seen as rational attempts by excluded groups to mobilize sufficient political leverage to advance collective interests through noninstitutionalized means. Sandweiss illustrates this theoretical framework by placing the environmental justice movement within the larger conceptual context of social action movements. He takes the view that the mobilization of activists and the securing of an official government response to the demands for environmental justice can be attributed, to a considerable degree, to the ability of the environmental movement to tap into the potent collective action frame of the civil rights movement. In the policy debate, grassroots movements by people of color and low-income groups must be shaped by the moral and rhetorical strength of civil rights and justice. Political solutions must be stressed over legal or scientific ones.

"Part 2: Environmental Injustices" pays closer attention to the factors contributing to the causes and consequences of environmental inequities. Using empirical evidence, Harvey L. White shows that not only are people of color endangered from environmental hazards more often, but that their level of endangerment is greater. The complexity of this phenomenon is underscored by the fact that in an attempt to improve their economic condition, people of color and low-income groups have directed their efforts toward bringing jobs to their communities even if these jobs pose great risks to the health of workers and the surrounding communities. The promise of jobs (even low-paying, hazardous ones) has enticed several economically impoverished, politically powerless communities. Jeanne Nienaber Clarke and Andrea K. Gerlak, through the case study approach, provide students with a further indication of how the political process model actually works. They also highlight two features of contemporary environmental politics. First, they show that people of color and the Anglo majority have distinct views on what is important, critical, and worth acting on around the issue of "environmentalism." The wider these distinctions, the more successful insurgency within and between these groups is undermined. Second, Clarke and Gerlak conclude that race and culture matter in the analysis of political representation. They raise issues around the cultural determination of attitudes about environmental injustices as well as perceptions about the nature of risk. Clarke and Gerlak make an all-important point: economics and race are inextricably intertwined. Kate A. Berry complements the notion of the "cultural determination of attitudes toward the environment" raised by Clarke and Gerlak. Her focus is on cultural values, particularly those espoused by Eurocentrism, as a determinant of public policies that adversely affect Native Americans. The legal doctrine that reinforces the Eurocentric value system is examined. Berry clarifies the value-laden character of public policy and shows that public policy is a political act, serving some actor's interests and promoting that actor's value judgments. She reinforces a point raised earlier by Clarke and Gerlak regarding political representation: the decision-making process reflects the values and interests represented in the existing regime and policy process. C. Richard Bath, Janet M. Tanski, and Roberto E. Villarreal describe the historical patterns of immigration leading to the *colonias* of El Paso, Texas. The economic aspects of the development of the *colonias* are discussed. This review brings us face to face with the question, Was the failure to extend the lines due to race or economics? This question is difficult to answer for Bath, Tanski, and Villarreal because of the subtleness of historically defined institutional discriminatory practices. It is difficult to determine "intent." Here, the intersection of race and class should be considered. Taken together, these four articles speak to the historical, political, economic, and cultural aspects of environmental injustices. This interdisciplinary approach complements the political analysis at the center of our investigation.

"Part 3: Confronting Environmental Injustices" addresses possible ways of dealing with the environmental problems disproportionately affecting people of color and low socioeconomic groups. The political process model identifies three sets of factors crucial to the generation of social insurgency. The first is the political alignment of groups within the larger political environment; the second, the level of organization within the aggrieved population; and third, the collective assessment of the prospects for successful insurgency within that same population. The two articles in this section examine some of these factors. John G. Bretting and Diane-Michele Prindeville look at grassroots activism as a form of political expression. Specifically, they look at the leadership role taken by women of color in developing movements meant to address environmental problems in New Mexico. They discuss the commonalities and differences between these indigenous women leaders, and the approaches to organizing that they utilized. Bretting and Prindeville shed some light on another important question: What is the relationship between environment, race, and gender? The differences between community- and nationally based environmental organizations, the motivational basis of leadership, the consequences of this leadership, and related issues merit attention in this original research. Peter J. Longo proposes a political strategy. He discusses the potential for coalition building between national environmental interest groups and groups formed to combat environmental injustices at the local level. Longo recommends that national environmental groups inject the "humanity" advocated by local groups into the legal formulas debated at the national level.

"Part 4: Environmental Justice" takes the view that it is irrelevant whether environmental injustice represents conscious racism or classism on the part of policymakers, either now or in the past. What really matters is how the problems are addressed by policymakers and business interests in the present. How these problems are addressed is of a political nature. Mary M. Timney suggests that providing the right incentives (negative or positive) can lead polluters to do what is best for society as well as their balance sheets. Requiring companies and subnational governments to focus on environmental justice as a goal may be the incentive for them to rethink the way they do business. But "politics as usual" must change. Policymakers must make a conscious effort to incorporate environmental justice into the "authoritative values for society." Linda Robyn and I also take an alternative "frame of reference" by pointing out that the Native American view of the environment is not taken into account when reaching public policy decisions. We challenge the dominant ideological framework utilized by decision makers in reaching conclusions about the environment, directly attacking "scientific" thinking. The last essay attempts to synthesize discussions raised throughout the book. In the final analysis, individual behavior must be grounded in the doctrines and principles of environmental ethics. As a form of applied normative ethics, environmental ethics deals with the approach to the environment that *ought* to be taken as well as the approach that *is* taken. My article demands critical thinking about the policies of the private and public sectors developed in response to environmental issues. I question the assumptions on which these policies are based and advocate new ways of thinking about the environment.

I

A Framework for Analysis

David E. Camacho

THE ENVIRONMENTAL JUSTICE MOVEMENT

A Political Framework

Individuals and groups in the United States have promoted protection of the environment since at least the 1830s. Two often conflicting concerns dominated the early environmental movement: (1) natural resources conservation, and (2) wilderness and wildlife preservation. One allowed for resource extraction and consumption; the other attempted to prevent such practices. At the same time, both concerns converged in their interest in the management of large, sparsely populated, public wetlands. From the 1950s onward, a third concern has influenced the modern environmental movement: human welfare ecology. Today, human safety and, in fact, survival are threatened by the dangers of nuclear plants, the growth in the nuclear arsenal, and nuclear wastes; by global warming and the thinning of the ozone layer; and generally, by ground, air, and water pollution. This threat to human survival is of a global dimension.[1] It was on the basis of these concerns that the environmental movement emerged as a significant mass movement in the 1970s. This modern, mainstream environmental movement has tended to exclude substantive participation by people of color, although people of color have organized around environmental issues at an unprecedented rate since the late 1980s.

Like many individuals and groups attracted to the contemporary environmental movement, people of color and low-income groups were horrified when they learned of the dangers to their communities caused by acute and chronic exposures to toxins and other environmental hazards. As noted by Peter J. Longo (this volume), however, mainstream environmental groups have still been slow in broadening their base to include people of color as well as poor and working-class whites. Like Longo, Stephen Sandweiss suggests that mainstream groups are ill equipped to deal with the environmental, economic, and social concerns of minority communities. For example, during the 1960s and 1970s, while mainstream environmental groups focused on wilderness preservation and conservation through litigation, political lobbying, and technical evaluation, people of color were engaged in collective action mobilizations for basic civil rights in the areas of employment, housing, education, and health care. Thus, two often conflicting movements emerged, and it has taken nearly two decades for any significant convergence to occur. While there are now efforts at coalition building, differences remain over how the two groups should balance economic development, social justice, and environmental protection. The crux of the problem is that the mainstream environmental movement has not fully recognized the fact that social inequality and imbalances of power contribute to the environmental degradation, resource depletion, pollution, and environmental hazards that disproportionately impact people of color along with poor and working-class whites. There is generally a lack of concern over "justice" in the mainstream environmental movement.

The environmental justice movement, then, is an attempt to unite the concerns of both the environmental and civil rights movements. Mainstream environmental organizations are beginning to understand the need for environmental justice and are increasingly supporting grassroots groups in the form of technical advice, expert testimony, direct financial assistance, fund-raising, research, and legal assistance. John G. Bretting and Diane-Michele Prindeville (this volume) point out that environmental justice advocates are not saying, "Take the poisons out of our community and put them in another community." They are saying that *no* community should have to live with these poisons. Bretting and Prindeville suggest that environmental advocates have taken the moral high road and that they are building a multiracial and inclusive movement that has the potential to transform the political landscape of this nation if we learn from its leadership style and organizational structure. The environmental justice movement is an inclusive one in that environmental concerns are not treated as separate and apart from health, employment, housing, and education issues. Indeed, Linda Robyn suggests that environmental justice cannot be separated from one's cultural lifestyle.

But what is to be done? Harvey L. White provides ample evidence that the record of environmental policy is not encouraging. The federal goverment has traditionally accepted major responsibility for protecting the health and well-being of U.S. society. During the 1980s, an extreme shift in governmental responsibilities emerged. This "New Federalism," ushered in by the Reagan administration, signaled a reduction in domestic programs to monitor the environment and protect public health. The shifting of environmental protection away from the federal government to subnational governments continues today. Actions by states are viewed as preferable and more effective. But the policy of delegating more responsibility for environmental protection and management to state agencies has serious implications. States generally do not have the expertise to handle this responsibility, and there has been no corresponding increase in financial resources to assist them in meeting these new responsibilities. The federal policy to continue with "New Federalism" has resulted in an abdication, rather than a shifting, of environmental protection and management.

Robert Bullard has provided "a glimpse of several representative struggles within the grassroots environmental justice movement."[2] Bullard offers insights into the coalescing of mainstream and grassroots environmental efforts, wherein he suggests that the situation is changing for the better. In addition, local governments have been moved to action and, in some cases, have also elicited the cooperation of environmental industries. But numerous challenges still face the environmental justice movement and its efforts to confront environmental problems. Sustaining a coalition based on racial, gender, and class differences is a formidable task; limited resources impede organizing; lack of information can block mobilization efforts; determining accountability and responsibility for environmental hazards can be impossible—these are some of the difficulties confronting the movement. In the end, they are political challenges.

Citizen deliberation is essential if there is to be a collective effort addressing environmental problems and solutions. Because the environmental crisis is a political one, thoughts on the environment must be cast in political terms. To this end, this chapter reviews the political process model. The intent is to identify factors that may lead to a successful insurgency around environmental injustices. Although the contributors to this volume will not necessarily make deliberate, direct references to the nature of power or the char-

direct references to the nature of power or the characteristics of politics as discussed below, as the reader continues through *Environmental Injustices, Political Struggles,* it is important that the framework discussed in this chapter guide one's critical analysis of the essays collected here.

The Political Process Model: Central Concepts and Assumptions

Politics refers to that set of social structures and processes by which humans resolve their conflicting interests without having to be in a constant Hobbesian "state of war." The term *process* conveys a main idea of the theoretical perspective taken in this anthology: a political movement represents a continuous process from generation to decline, rather than a discrete series of developmental stages. Accordingly, the political process model offers the reader a framework for analyzing the entire process of a movement's development rather than a particular phase (e.g., the emergence of the tactic of protest) of that same process. At its most elementary level, the study of politics is concerned with the processes that determine "who gets what, when, and how."[3] David Easton suggests that politics is the "authoritative allocation of values for society."[4] These two views of politics provide insights into the study of power. Power is "the ability of A to somehow affect the behavior of B," suggesting that at its most basic level, power can be seen as the capacity of some groups or individuals to impose their preferences on others within a collective.[5]

The Lasswellian view breaks politics into four related parts: (1) who acts, (2) what is being sought, (3) when (the empirical context), and (4) how (the process). This perspective identifies the actors who pursue the rewards and benefits of political participation. The attention given to the political process helps answer significant questions: How do actors accomplish their political objectives? How do actors lose out in their pursuit of rewards? Further, it helps identify competition among actors as well as the historical development of that competition. The notion of historical development is an important one. As regards the environmental justice movement, it suggests that the tenets of the movement cannot be separated from the past socioeconomic and political conditions that spurred it.

Easton points out that *order* is a necessary condition of society. As such, people within a political system authorize government (or its public officials) to manage conflict in society. Government, then, is given legitimacy to manage the affairs of its citizens through the "social contract" entered into by a political community. The management of conflict is accomplished primarily through a system of rewards and punishments, or by rules that provide an orderly manner by which groups and individuals can interact with each other. Witness, for example, the valuable guidance offered by constitutions in the U.S. political system. Government, in its provision of social order, is responsible for making sure that human behavior is kept within the parameters of "the accepted way of doing things."

An important way that government maintains the social order is by its provision of collective goods and services: it supplies those rewards and benefits demanded or needed by the public at large and, in some instances, by private individuals. This capacity to provide collective goods and services also implies that government has the capacity to withhold rewards and benefits. Excluding some groups and individuals becomes especially necessary under the condition of "limited resources." A condition of limited resources, like the economic condition of "scarcity," exists when wants exceed available goods and services. Government simply does not have *all* the resources at its disposal; thus, the role of government in managing the political process becomes problematical because wants inevitably exceed the available goods and services that government possesses.

This suggests that the role of government is to carry out two necessarily crucial functions for society: (1) the provision of social order, and (2) the provision of public goods and services. These functions are not mutually exclusive. Used in combination, they reinforce each other in effectively meeting the ends of the state: the provision of goods and services essential to maintain the social order. Government's role becomes critically important, then, under the condition of limited resources. Indeed, government exists to decide (authoritatively) how limited resources (values) are to be allocated for society so that order is maintained. Since there is not enough of what people want for everyone, each public policy decision requires that some groups and individuals make sacrifices. In determining the winners and losers involved in the competition over limited resources, government takes on a special significance because it is dictating power relations in society.

Government's sanctioning of power relations in society is, therefore, a result of its role in managing conflict in society. Government upholds, justifies, and legitimates traditional modes of behavior and an established pattern of authority in assuring that human behavior is kept within the parameters of the accepted way of doing things. In the end, public policy decisions aimed at resolving political conflicts over resource allocations, procedural rules and regulations, and values will mostly favor these traditional and established patterns. Consequently, both human behavior and political action are significantly shaped by the institutional and procedural parameters of a society. Stated differently, a key factor in understanding political development and behavior is the institutional framework of U.S. society. For example, individuals engaged in mainstream politics are not likely to upset established power relations because their political behavior is within the parameters of the accepted way of doing things; conversely, protest is a type of political behavior that falls outside of mainstream politics and can be understood as a reaction to glaring inequalities or a lack of political resources.

In brief, then, politics involves power when it is viewed as a process whereby groups or individuals compete against each other over limited resources, and as a result, there are winners and losers. Conflict is a fundamental condition of politics. Power relationships exist when there is conflict over resource allocations, procedural rules and regulations, and values.

The Political Process Model: Theoretical Perspective

The political process model is based on an elitist conception of power.[6] Power is conceptualized as the ability of A to prevail over B in formal decision making; A also prevails over B in determining what is to be deemed a formal issue. Elite theory argues that there are essentially two major groups in society: a small group of powerful elites and the powerless masses. Under elite theory, people of color and the working poor are part of the powerless mass. The elite group is well organized and controls the major economic, social, and political institutions in society. While the masses might influence government through elections, their influence is seen as marginal rather than as a serious challenge to elite dominance, which is embedded in socioeconomic institutions and reinforced by ideology, or a dominant set of values, beliefs, attitudes, and norms. Social movements are viewed as attempts by excluded groups to mobilize sufficient political leverage to influence elite dominance.

Where our perspective diverges from the elite model is in regard to the extent of elite control over the political system and the insurgent capabilities of excluded groups. Theorists of the elite model tend to hold a view of the disparity between elite and excluded groups that grants the former virtually unlimited power. Excluded groups are seen as functionally powerless compared to the enormous power wielded by the elite. Under such conditions, the chances for successful insurgency would be nil. The power disparity between elite and excluded groups is substantial, but there is potential for insurgency among excluded groups. Political efficacy is crucial for awakening latent insurgency. That is, members of excluded groups must believe in their ability to mobilize significant political resources and in their skill to effectively exercise political leverage (see Bretting and Prindeville, this volume).

The political process model is consistent with Easton's views on "insiders" and "outsiders" and the conservatism of the political system. According to Easton, political actors inside a political system hold the values of that system and perceive its interests as their own. Political interaction within the system involves disagreements, to be sure, but within the bounds of a basic consensus or within the parameters of "the accepted ways of doing things." Political actors outside a political system typically will not share certain basic values with the groups and authorities inside the system. The basic values held by insiders and outsiders could be so fundamentally different that a peaceful coexistence is impossible (see, this volume, Clarke and Gerlak; Berry; Robyn and Camacho). The interaction between excluded groups and the system would involve conflict to the extent that the relationship would revolve around a clash of values. Like Easton's thinking, the political process model is based on the notion that political action by insiders reflects an abiding conservatism. Accordingly, insiders resist changes that would threaten the realization of their interests, fight tenaciously against a loss of power, and work to bar admission to the political system of political actors whose interests conflict significantly with their own.

This conservatism brings to bear sociological arguments regarding the nature and importance of social stratification to better understand power relationships.[7] The resulting differences of socioeconomic status or lack of it have clear and major implications for the distribution of power in the political and social system. Government officials depend heavily on upper-class groups or interests because they possess most of the resources necessary to advance governmental activities and policies. Not only do upper-strata interests have a greater quantity of resources, they also have a greater range of resources—more durable and more indispensable—than others. Some upper-class groups, including large businesses and corporations, hold what is essentially public power to the extent that upper-strata interests are a factor in all that public officials do.

The importance of this sociological view for understanding the environmental justice movement is the attempt to link issues of class with those of racial politics. A group's social-class position serves as a clear indication of its attractiveness as an insider or outsider. "Minority status"—held by members of the environmental justice movement—connotes an image of a group that is undesirable, undeserving, and probably has little to offer politically. People of color and low-income groups are perceived as less desirable than upper-class groups. The upper-strata (or "winners") prevail consistently though not totally because, in part, they enjoy a systemic advantage in the governmental process. Over time, "losers" suffer a cumulative systemic disadvantage and become outsiders to the governmental process. In the end, a relationship between class and race exists, and "minority" political opportunities and influence are especially limited by that relationship.

The strategic constraints confronting excluded groups should not be underestimated. But if elite groups are unwilling to support insurgency, the very occurrence of social movements indicates that indigenous groups are able to generate and sustain organized mass action. The political process model assumes that social movements develop in response to an ongoing process of favorable interplay between movement groups (internal factors) and the larger sociopolitical system (external factor) they seek to change. The specific combination of factors may change from one phase of the movement to another, but the basic interaction between the two remains the same.

Factors for Successful Insurgency

The political process model identifies three factors crucial to the generation of social insurgency: (1) the political opportunities afforded excluded groups within the larger political environment, (2) the level of organization within the excluded group, and (3) the collective assessment of the prospects for successful insurgency within that same group. Each of these factors is discussed below. Sandweiss, in the next chapter, complements and adds insight into this discussion. For instance, he evaluates the relationship of the environmental justice movement to the larger political environment—Congress, the White House, the Environmental Protection Agency—in an attempt to assess the political opportunities afforded the movement by those governmental bodies. From Sandweiss's discussion, the reader is asked to consider the power relations sanctioned by government. Whose values and interests are given authoritative dominance?

POLITICAL OPPORTUNITIES

As noted above, under ordinary circumstances, excluded groups or outsiders face enormous obstacles in their efforts to advance group interests. Outsiders are excluded from decision-making processes precisely because their bargaining position is so weak compared to the power of insiders. But the system will be more or less open to specific groups at different times and at different places. The opportunities for an excluded group to engage in successful collective action can occur. What accounts for political opportunities? Peter Eisinger suggests that "shifts"—any event or broad social process that serves to undermine the fundamental ideas and assumptions on which the political establishment is structured—in the political system itself provide these opportunities.[8] Among the events and processes likely to prove disruptive of the status quo are wars, industrialization, urbanization, prolonged unemployment, and demographic changes. In combination, shifts place extreme pressure on the status quo, as when great numbers of southern Blacks moved to the industrial centers of the urban East Coast during World War II. The political process model views these shifts as promoting movement insurgency indirectly through a restructuring of existing power relations. Moreover, social insurgency is shaped by broad social processes that operate over a long period of time. The processes shaping insurgency are of a cumulative nature.

In short, instability brought on by shifts in the political, social, and economic landscapes disrupt the status quo and thus encourage collective action by groups sufficiently organized to contest the structuring of a new political order. This does not mean that the interests of the excluded group will be realized, for even in the context of an improved bargaining position, excluded groups still have a distinct disadvantage in any confrontation with an established political actor. What this does mean is that the increased political leverage of the aggrieved community—those experiencing glaring inequalities and lacking mainstream political resources—has improved the bargaining position of insurgent groups and, hence, creates new opportunities for the collective pursuit of group goals. It means, too, that the improved bargaining position of the excluded group significantly raises the cost of repressing insurgent action. Unlike before, when the powerless status of the excluded group meant that it could be repressed with relative ease, the increased legitimacy garnered by the insurgent group now renders it less vulnerable. Increased political power, legitimacy, and a sense of efficacy serve to encourage collective action by diminishing the risks associated with movement participation.

The Browning, Marshall, and Tabb study is instructive here.[9] Their study is guided by three central questions: (1) How open are urban political systems? (2) How does political incorporation occur? (3) Does the political incorporation of people of color make a difference for policy? They identify levels of political development and behavior: protest demands, electoral mobilization, representation, and incorporation. They find that protest does not adequately bring about policy responsiveness. What is decisive is incorporation, or the substantive participation by African Americans and Latinos in liberal coalitions in city government. Incorporation depends heavily on electoral mobilization, which, in turn, is often preceded by protest. Policy responsiveness is largely determined by levels of incorporation.

While their findings inform the understanding of racial politics in our cities, the approach of the study better serves our purposes. Browning, Marshall, and Tabb's study takes as a given the strains of urbanization on people of color; it reflects the changes resulting from demographic shifts; longitudinal data stress the cumulative effect of racial mobilization efforts; and it suggests a process facilitated by political instability (specifically, the social unrest of the 1960s) that might lead to incorporation.

LEVEL OF ORGANIZATION

The second factor can be understood as the degree of organizational readiness within the aggrieved community. A favorable external environment can afford the aggrieved community the opportunity for successful insurgent action; however, its resources enable insurgent groups to exploit these opportunities. In the absence of those resources, the aggrieved group will likely lack the capacity to act even when granted the opportunity to do so. This view is compatible with the resource mobilization perspective of pluralism.[10] Pluralism assumes that there are multiple centers of power, not a single center, in society and politics. Power is considered to be widely dispersed among a variety of groups and institutions in society. These groups have some degree of essential political resources, including group size, financial resources, social status (prestige and legitimacy), cohesiveness, effective leadership, political knowledge, and intensity. Pluralism asserts that these resources are noncumulative; thus, groups advantaged in one resource (e.g., intensity) may be weak in another resource (say, legitimacy). Political tactics stem from available resources: a financially poor but numerically large group may stress electoral politics over campaign financing in its attempt to influence the political process. One group may have more of a particular resource than another, but overall resource equality is maintained through the noncumulative pattern of group resources.

Pluralism has its critics.[11] For example, pluralism argues that if a group is discriminated against, the discrimination is at least partly attributable to the group's inability to effectively use its political resources. Certain legal protections are supposed to ensure that all groups, including racial and low-income groups, have the rights of free association, free speech, and so on. Those rights are essential to the realization of civil liberties, and they limit the existence and degree of discrimination. Critics argue that such protections are necessary but not sufficient to overcome substantial political, economic, and social inequalities and their consequences. Another assumption of pluralism is that there are multiple access points in the political system. A group that has been denied access in its efforts to influence politics at one point in the system can pursue its political interests in other places or arenas. If a group has been unable to influence the electoral process, it may seek judicial redress; or if local government proves to be unresponsive to group needs, the group may seek to influence the state or federal government. This assumption is highly problematic.[12] It views groups as mobilized at all levels of government and with enough resources to redirect their mobilization efforts from one arena to another.

These criticisms are well taken, but pluralism does provide insight into the extent of organization necessary for the aggrieved group to generate a social movement. Bretting and Prindeville (this volume) suggest that indigenous structures frequently provide the organizational base out of which social movements emerge. Sandweiss (also this volume) stresses the importance of an established network in the generation of social insurgency. He argues convincingly that the ability of those adversely affected by environmental injustices to generate a social movement is partly dependent on the presence of an indigenous "master frame" that can be used to link members of the aggrieved population into an organized campaign of mass political action. Through a "frame extension," the environmental justice movement has been attempting to link itself to "the rhetorical legacy of the civil rights movement," according to Sandweiss, and its considerable organizational resources and communication networks.

Established networks or organizations provide four crucial resources to excluded groups.

Members

Sandweiss points out that if there is a consistent finding in the empirical literature, it is that movement participants are recruited along established lines of "articulation," including areas of problem definition, causal attribution, moral evaluation, and remedial action. The significance of this finding is clear: the more integrated the person is in the aggrieved community, the more readily he or she can be mobilized for participation in protest activities. A characteristic of the indigenous leaders evaluated by Bretting and Prindeville is the leaders' integration into the communities they represent. The relevance of grassroots organizations stems from the fact that they render this type of integration more likely, thus promoting member recruitment. Accordingly, individuals can be recruited into the ranks of movement activists by virtue of their involvement in organizations that serve as a network, out of which a new movement emerges. This is true, as Sandweiss notes, in the case of the environmental justice movement, with a disproportionate number of the movement's recruits coming from existing civil rights groups.

Grassroots organizations, moreover, can serve as the primary source of movement participants, wherein movements do not so much emerge out of established organizations as they represent a merger of such groups. Sandweiss describes how the environmental justice movement emphasized such values as individual rights, equal opportunities, social justice, citizenship, and human dignity, thus legitimating the struggles of other disenfranchised groups. The environmental justice movement has merged with civil rights concerns around equal education, employment, housing, and health care.

Mobilization does not occur through recruitment of large numbers of isolated and solitary individuals. Rather, it is a result of recruiting prospective members who are already highly organized and active.

Incentives for Solidarity

What explains an individual's decision to join a movement? First discussed by Mancur Olson, the rational individual could take a "free ride"—that is, choose not to participate in the movement and still benefit from the efforts of others.[13] When viewed from an economic cost-benefit analysis, movement participation would indeed seem to be irrational. The free-rider mentality poses a formidable barrier to movement recruitment. The solution to this problem is held to stem from the provision of selective incentives to induce the participation that individual cost-benefit analysis would alone seem to preclude.[14] Selective incentives are direct member inducements, which the individual receives in exchange for his or her contribution to and participation in the movement.

In the context of existent organizations, however, the provision of selective incentives seems unnecessary. Established organizations already rest on a solid structure of solidary incentives. Movement leaders need only appropriate these incentives by defining movement participation as synonymous with organizational membership. Accordingly, the incentives that have served as the motive force for participation in the group are now simply transferred to the movement. Thus, movement leaders have been spared the difficult task of inducing participation through the provision of new incentives of either a material or nonmaterial nature. This phenomenon is clearly explained by Sandweiss in the appropriation of fundamental values and beliefs of established civil rights organizations by the environmental justice movement. Peter Longo (this volume) views this task as crucial to efforts at coalition-building.

A second resource available to an emerging social movement through established grassroots organizations, then, are the incentives already held and relied on by indigenous group members that encourage solidarity and participation.

Communication Network

The established organizations of the aggrieved community also constitute a communication network or infrastructure, the strength and breadth of which largely determine the pattern, speed, and extent of movement expansion. The development of the environmental justice movement highlights the salience of such a network precisely because the conditions for a movement existed *before* a network came into being, but the movement did not exist until afterward. From 1965 to 1985, for instance, socioeconomic strains and environmental hazards did not change significantly for people of color and low-income groups. Indeed, if anything, their condition deteriorated during this twenty-year period. What changed was the organizational situation. It was not until a communications network evolved among like-minded people beyond local boundaries that the movement could emerge and develop past the point of occasional, spontaneous uprisings.

Simply put: the interorganizational linkages characteristic of established groups facilitate movement emergence by providing the means of communication by which the movement can be disseminated throughout the aggrieved group.

Leaders

Common sense points to the salience of leaders or organizers in the generation of social insurgency. The complexity of the political game and the likely widespread discontent and conflict associated with politics necessitate the centralized direction and coordination of a recognized leadership. The existence of established organizations within the movement's mass base ensures the presence of recognized leaders who can be called on to lend their prestige and organizing skills to the incipient movement. In fact, given the pattern of communication discussed in the previous section, it may well be that established leaders are among the first to join a new movement by virtue of their central position within the aggrieved community. According to this vein of thought, no wonder it was the Commission for Racial Justice of the United Church of Christ that published the most widely recognized study of race and the incidence of environmental hazards in the United States. Regardless of the timing of their recruitment, the existence of recognized leaders is yet another resource whose availability is conditioned by the degree of organization within the excluded group.

In the final analysis, established organizations within the excluded community are the primary source of resources facilitating movement emergence. These indigenous, grassroots groups constitute the organizational context in which movement insurgency develops. In the absence of this supportive, facilitative organizational context, the aggrieved group is likely to be deprived of the capacity for collective action even when given the opportunity to participate by insiders. If a political actor lacks the capacity to act, it hardly matters that an opportunity to participate has been provided.

ASSESSMENT OF THE PROSPECTS FOR SUCCESS

Political opportunities afforded by political, social, and economic shifts or by established indigenous organizations do not manufacture a social movement. These two factors are necessary, but insufficient, causes of insurgency. Together, they only offer the "potential" for collective action. Intervening between the "objective" factors of political opportunities and resource mobilization are *people* and the "subjective" meanings they attach to their situation. It is generally held that individuals react to the political system according to how that system has treated them. As Charles V. Hamilton explains, "People who have experienced positive results (albeit limited in many instances) as a function of mobilization and bargaining are much more likely to have respect for that process—indeed, to participate in it—than those who have been thwarted at trying to enter that process."[15]

In other words, will favorable shifts in political opportunities be defined as such by a large enough group of people to facilitate collective protest? This assessment of the prospect for success is not independent of the two factors discussed previously. One effect of improved political opportunities and established organizations is to render this assessment of "efficacy" more likely. I explore the relationship between this process of assessment and each of these two factors separately.

As noted above, favorable shifts in political opportunities decrease the power disparity between excluded groups and insiders, and in doing so, increase the cost of repressing the movement. These are objective structural shifts. Such shifts, however, induce changes of a subjective nature as well. Murray Edelman has pointed out that "cues" are crucial in shaping the subjective meanings individuals attach to their situations.[16] Edelman would suggest that outsiders experience shifting political opportunities on a day-to-day basis as a set of "meaningful" events communicating much about their prospects for successful collective action. Stated similarly, Hamilton writes:

People are politicized frequently by their real experiences. One relates to what I shall characterize as the 3-P proposition: Process, Product, Participation.

The proposition is a reasonably simple one: to the extent that the *process* is perceived as related to the *product* desired, then *participation* will increase. There is nothing complicated about this. If people believe that if they vote or march or write letters or strike (the *process*), they will very likely get good housing or clean streets or good schools (the *product*), they will then *participate* in the particular endeavor.

The converse is clear: if certain processes are deemed irrelevant or ineffective, they will be rejected. Therefore, when we see low voter turnout or low participation in other activities, we should not be too quick to conclude apathy. An apathetic person is not necessarily alienated; he might simply be overburdened by personal concerns. But this is not to be confused with an attitude of alienation, which is far more judgmental. The alienated person does not participate because he or she has given up on the processes offered. He sees them as altogether unlikely to yield a desired product. Such a person is still concerned, but sees no *viable* means for effectuating that concern. Such a person is far more likely to be a recruit for destructive deviant activity than one who is simply apathetic, or if appealed to convincingly, a recruit for activity in behalf of high ideals.[17]

It appears, then, that this "3-P proposition" must occur if an organized protest campaign is to take place. Sometimes, the political importance of events is clearly apparent, as when mass migration significantly alters the electoral composition of a region.[18] But even when evolving political meanings are of a less overt nature, they will invariably be made available to individuals through subtle cues communicated by other groups. Here, the Browning, Marshall, and Tabb study is again instructive. In a highly competitive electoral arena, we should expect local officials to be more responsive to African American and Latino voters than they have previously been.

As overt or subtle as these cues may be, their significance for the generation of insurgency cannot be overstated. The accommodating responses of insiders to a particular outsider serve to transform evolving political opportunities into a set of meaningful events or cues signifying to members of excluded groups that the political system is becoming increasingly open to their demands. Thus, by forcing a change in the substantive or symbolic content of insider/outsider power relations, shifting political opportunities supply a crucial catalyst in the process of efficacy.

Indigenous, grassroots organizations in the aggrieved community also have a pronounced effect on the development of political efficacy. The relevance of established organizations to social movements was noted earlier, and the argument advanced was that established organizations afforded incipient social movements a communication network ensuring the thorough dissemination of social insurgency throughout the aggrieved community. That insight can now be extended further. It is not simply the extent to which—or speed with which—insurgency is spread that is important here. Rather, the cues and meanings on which insurgency depends are shaped by the strength of integrative ties within the movement's mass base. That is, the process of efficacy is held to be more likely and of far greater consequence under conditions of strong social integration. Conversely, the absence of integrative links would probably deter the spread of efficacy to enough individuals to sustain a reasonable chance for successful collective action. Indeed, lacking the information and cues that others provide, isolated individuals may explain their condition on the basis of personal rather than systemic deficiencies. The significance of this allusion to self-blame is the fact that only systemic conditions furnish the necessary rationale for movement activity. The consistent finding linking feelings of political efficacy to social integration supports this view (see Bretting and Prindeville, this volume).

To summarize, movement emergence requires a transformation of consciousness (feelings of efficacy) within a large segment of the aggrieved community. Before the protest stage can begin, individuals must collectively define their situation as unjust and vulnerable to change through group action. The prospect of this transformation occurring is determined by the structure of opportunities and the existence of indigenous organizations. Shifting political, social, and economic conditions supply the necessary "cues" capable of awakening the process of political efficacy. Existent organizations give the members of the emerging social movement both a communications network and the stability of an integrative group setting within which efficacy is most likely to occur.

Conclusion

The inception of a social movement presupposes the existence of a political environment increasingly vulnerable to pressure from insurgents. Specific events and shifting social processes enhance the bargaining position of the aggrieved community. Organizational readiness augments insurgent groups mobilized to exploit the expanding opportunities for collective action. The survival of a social movement requires that insurgents be able to maintain and successfully utilize their newly acquired political leverage to advance collective interests. If they are able to do so, the movement is likely to survive. If, on the other hand, insurgent groups fail to maintain a favorable bargaining position vis-à-vis other groups in the political arena, the movement faces extinction. In short, the ongoing exercise of significant political leverage remains key to the successful development of the movement.

What is missing from the above discussion is any recognition of the immense countermobilization insurgents must overcome if they are to succeed in their collective action efforts. Even as a social movement exploits political opportunities, it sets in motion processes contrary to and possibly destructive of insurgency. Because of the political system's bias toward mobilization efforts by formally recognized organizations, leaders must be able to create a more enduring organizational structure for the movement to survive. These efforts usually entail the creation of formal organizations to assume a centralized direction for the movement, which was previously directed by informal groups. This transformation occurs if the resources needed to generate the development of the movement's formal organizational structure can be mobilized. Accordingly, insurgent leaders must be able to exploit the initial successes of the movement to mobilize those resources needed to facilitate the development of the more permanent organizational structure required to sustain insurgency. Failing this, movements are likely to die as the loosely structured groups previously guiding the protest campaign disband or gradually lapse into inactivity.

Notes

1. See, for example, Jeremy Rifkin, *Biosphere Politics: A Cultural Odyssey from the Middle Ages to the New Age* (San Francisco: Harper Collins, 1991).

2. Robert D. Bullard, ed., *Confronting Environmental Racism: Voices from the Grassroots* (Boston: South End Press, 1993), 15.

3. Harold D. Lasswell, *Politics: Who Gets What, When, and How?* (New York: World Publishing, 1958).

4. David Easton, *The Political System* (New York: Knopf, 1953).

5. Steven Lukes, *Power: A Radical View* (London: Macmillan Education, 1989).

6. See, for example, C. Wright Mills, *The Power Elite* (New York: Oxford University Press, 1956); Joseph Schumpeter, *Capitalism, Socialism, and Democracy* (New York: Holt, Rinehart, and Winston, 1960); Peter Bachrach and Morton S. Baratz, "Two Faces of Power," *American Political Science Review* 56 (1962): 947–52; and Roger W. Cobb and Charles D. Elder, *Participation in American Politics: The Dynamics of Agenda-Building,* 2d ed. (Baltimore, Md.: Johns Hopkins University Press, 1983).

7. Clarence N. Stone, "Race, Power, and Political Change," in *The Egalitarian City,* ed. Janet K. Boles (New York: Praeger, 1986), 200–223.

8. Peter Eisinger, *The Politics of Displacement: Racial and Ethnic Transition in Three American Cities* (New York: Academic Press, 1980).

9. Rufus P. Browning, Dale Rogers Marshall, and David H. Tabb, *Protest Is Not Enough: The Struggle of Blacks and Hispanics for Equality in Urban Politics* (Berkeley: University of California Press, 1984).

10. Robert A. Dahl, *Who Governs?* (New Haven, Conn.: Yale University Press, 1961).

11. A basic, applied criticism is John Harrigan, *Empty Dreams, Empty Pockets: Class and Bias in American Politics* (New York: Macmillan, 1993).

12. See, for example, Rodney E. Hero, *Latinos and the U.S. Political System: Two-Tiered Pluralism* (Philadelphia: Temple University Press, 1992).

13. Mancur Olson, *The Logic of Collective Action: Public Goods and the Theory of Groups* (Cambridge, Mass.: Harvard University Press, 1965).

14. Terry M. Moe, *The Organization of Interests: Incentives and the Internal Dynamics of Political Interest Groups* (Chicago: University of Chicago Press, 1980).

15. Charles V. Hamilton, "Political Access, Minority Participation, and the New Normalcy," in *Minority Report: What Has Happened to Blacks, Hispanics, American Indians, and Other Minorities in the Eighties,* ed. Leslie W. Dunbar (New York: Pantheon, 1984), 11.

16. Murray Edelman, *The Symbolic Uses of Politics,* new ed. (Urbana: University of Illinois Press, 1985); and Edelman, *Constructing the Political Spectacle* (Chicago: University of Chicago Press, 1987).

17. Hamilton, "Political Access," 15–16.

18. Browning, Marshall, and Tabb, *Protest Is Not Enough.*

Stephen Sandweiss

THE SOCIAL CONSTRUCTION OF ENVIRONMENTAL JUSTICE

In September 1982, over 400 protesters were arrested at a proposed toxic waste landfill site in Warren County, North Carolina. At the time, the population of the county was 84 percent African American, and it was also one of the poorest counties in the state, with a median income of $6,984 and an unemployment rate of 13.3 percent. The protesters, including several leaders of national civil rights organizations, were convinced that these factors, and not the geological suitability of the disposal site, were responsible for the decision to bury 32,000 cubic yards of soil contaminated with highly toxic PCBs in Warren County.[1] Although the protest failed to prevent the opening of the landfill, it served to galvanize a nationwide grassroots social movement demanding "environmental justice."

Today, there are approximately 200 groups representing people of color around the country that are actively working on environmental issues, either as their principal focus or as part of a broader social justice agenda.[2] At the same time, concerns for equity and justice have come to occupy a distinct place on the environmental policy agenda, at both the federal and state levels.[3] Moreover, several members of Congress have introduced environmental justice bills and committee hearings have been held on the subject. Within the executive branch, the administrator of the Environmental Protection Agency (EPA), Carol Browner, has publicly announced that her agency is committed to incorporating environmental justice concerns into everything we do. To this end, the EPA established an Office of Environmental Justice and a National Environmental Justice Advisory Council, and has also undertaken an investigation into claims of environmental discrimination under Title VI of the 1964 Civil Rights Act. Most prominently, in February 1994, President Clinton issued Executive Order 12898, directing all federal agencies to take steps to ensure that minority and low-income communities are not disproportionately affected by environmental burdens.

This chapter explores two related questions: (1) Why has the environmental justice movement been successful in mobilizing activists and securing a place on a crowded political agenda? and (2) Will this success translate into substantive policy changes? To address these questions, I employ a social constructionist approach to the role of social movements in the policy process. Such an approach is based on the contention that social problems are not simply reflections of objective conditions in society but rather come to be identified as such only as a result of an interpretive process engaged in by competing claims makers.[4] This approach attempts to explain why only some social conditions come to be defined as problems. Central to this analysis is the concept of an interpretive frame, which not only calls attention to a problem but also seeks to identify both its causes and possible remedies. In the next section, I identify the conceptual components of a collective action frame and discuss the significance of movement frames in the policy process. Subsequently, I examine the general frame advanced by participants in the environmental justice movement. This essay asserts that the mobilization of activists and the securing of an official government response to the demands for environmental justice can be attributed, to a considerable degree, to the ability of the movement to tap into the potent collective action frame of the civil rights movement.

Following this, I compare the movement frame to those proferred by other key actors in the policy process: Congress, the White House, and the EPA. In so doing, I demonstrate that, although a relative consensus exists concerning the definition of the problem of environmental injustice, there remains considerable disagreement regarding its causes and possible solutions. This disagreement seriously threatens the ability of the environmental justice movement to achieve the substantive policy changes it is demanding. The chapter concludes by arguing that the environmental justice movement must be wary of getting caught up in the legal and highly technical discourse that characterizes much of the policy debate, and must instead continue to employ an aggressive, grassroots focus on political empowerment, voiced in the language of social and economic justice.

Collective Action Frames and Problem Definition

Over the past three decades, students of social movements have emphasized the presence of both sufficient material resources and favorable political opportunities as integral to the success of movement mobilization.[5] More recent scholarship, however, has stressed the significance of values, beliefs, ideas, and grievances in the emergence and subsequent life cycle of social movements. Accordingly, many theorists have come to recognize the merit of integrating a social constructionist perspective into their analyses of movement dynamics.[6] A key element in this constructionist approach is an emphasis on the development of a "collective action frame," which serves to articulate movement grievances, generate member solidarity, and motivate participants to action.

Before discussing the components of this collective action frame, it is useful to briefly define the concept of a frame. In an effort to clarify what he argues has become a "fractured paradigm," Robert Entman proposes the following general theory:

> Framing essentially involves selection and salience. To frame is to select some aspects of a perceived reality and make them more salient in a communicating text, in such a way as to promote a particular problem definition, causal interpretation, moral evaluation, and/or treatment recommendation for the item described.[7]

According to Snow and Benford, collective action frames fulfill two important functions in the emergence of social movements: punctuation and attribution. First, they explain, collective action frames permit activists to "single out some existing condition or aspect of social life and define it as unjust, intolerable, and deserving of corrective action."[8] Equally important, contend Snow and Benford, collective action frames make both diagnostic and prognostic attributions, by assigning blame or responsibility for the problematic condition and by suggesting appropriate remedies.

A fair amount has been written on the importance of movement frames as "internal" resources that groups use to generate activism. Recently, attention has been paid to the "external" role of frames as "tools that groups wield more or less self-consciously in their social and political struggles."[9] In an article on the various social constructions of the "public good" advanced by social movements, Rhy Williams argues that movement frames must be considered as strategic cultural resources, along the same lines as more conventional structural resources such as money, members, and organizational ties.

The way in which movement grievances are framed goes a long way toward determining the ability of movements to secure a place for their grievances on the public and policy agendas. That is, entrance into public political arenas requires legitimate cultural resources as a medium for power. One crucial resource, Williams suggests, is a vision of the "public good" that resonates with enduring values in American political culture.

> Movements use particular constructions of the public good in an attempt to frame public politics in ways that provide advantage for their agenda. By talking about the public good in a particular way, movements simultaneously legitimate their involvement and solutions, while casting aspersions on their opponents' positions.[10]

Despite the fact that it is not phrased in the same conceptual language, this call for a focus on the contextual and public dimensions of movement frames fits nicely with the emerging policy process literature in the area of problem definition. Heeding E. E. Schattschneider's dictum that "the definition of the alternatives is the supreme instrument of power,"[11] policy scholars have shown an increased interest in the social construction of political problems and their impact on the agenda-setting process.[12] Essentially, what these authors have demonstrated is that policy debates involve "framing contests," in which interested parties put forth various interpretations of political problems, as well as ideas about their causes and possible remedies, in hopes of placing an issue on—or keeping it off—the formal policy agenda.

The Environmental Justice Movement Frame

Given the dual role played by movement frames—mobilizing activists and defining a public position in the policy debate—the collective action frame of the environmental justice movement can be useful in helping us to understand why the movement has grown so successfully, and why equity and justice concerns have achieved prominence on the environmental policy agenda. Using the four frame elements suggested by Entman above, I summarize the various components of an environmental justice frame. I then argue that the mobilizing potential and rhetorical power of the frame stem from its ability to extend the "master frame"[13] provided by the civil rights movement.

PROBLEM DEFINITION

For most environmental justice activists, the problem is regarded as one of distributional inequity. Throughout the literature on environmental justice, one encounters the claim that "low-income and minority communities continue to bear greater health and environmental burdens, while the more affluent and whites receive the bulk of the benefits."[14] In support of this claim, members of the environmental justice movement point to a wide range of scientific research.

In a review of sixteen previous studies, Paul Mohai and Bunyan Bryant note that the findings "indicate clear and unequivocal class and racial biases in the distribution of environmental hazards."[15] Of these studies, the one most frequently mentioned by environmental justice activists is a 1987 report commissioned by the United Church of Christ entitled *Toxic Wastes and Race*. This was the first national study to analyze the demographic composition of communities surrounding commercial hazardous waste facilities and uncontrolled toxic waste sites. The report, using multivariate analysis to control for the effects of socioeconomic status, found that race was consistently a more prominent factor in the location of commercial hazardous waste facilities than any other factor examined.

Another study often cited to demonstrate the inequity of the distribution of environmental hazards is a 1992 investigation by the *National Law Journal* (*NLJ*) entitled "Unequal Protection: The Racial Divide in Environmental Law." This study examined EPA cleanup efforts at over 1,100 Superfund sites. The *NLJ* concluded that the average fine imposed on polluters in white areas was 506 percent higher than the average fine imposed in minority communities. It also discovered that cleanup took longer in minority communities, even though the efforts were often less intensive than those performed in white neighborhoods.[16]

CAUSAL ATTRIBUTION

The cause of these distributional inequities in both environmental planning and enforcement is asserted to be "environmental racism," a term coined by Dr. Benjamin Chavis in the aftermath of the 1982 protests at Warren County. According to Chavis, environmental racism refers to:

Racial discrimination in environmental policy making and the unequal enforcement of the environmental laws and regulations. It is the deliberate targeting of people-of-color communities for toxic waste facilities and the official sanctioning of a life-threatening presence of poisons and pollutants in people-of-color communities. It is also manifested in the history of excluding people of color from the leadership of the environmental movement.[17]

By employing the word "deliberate," Chavis is suggesting that the causes of environmental discrimination are intentional. Not all movement members, however, consider intent as the standard for defining environmental racism. According to Robert Bullard, a sociologist who is by far the leading academic activist on environmental justice issues, environmental racism refers to

any policy, practice, or directive that differentially affects or disadvantages (whether intended or unintended) individuals, groups, or communities based on race or color.[18]

By implying that one need not prove intent in order to demonstrate the existence of environmental racism, Bullard is shifting the standard of proof to one of disparate impact or effects. Equally important, Bullard and many other environmental justice advocates point to the "institutionalized" nature of environmental racism, linking it to other forms of racial discrimination that mirror the power arrangements of the dominant society. For those who subscribe to this view, segregation in both the residential and labor markets is what "enables environmental burdens to be inequitably distributed in the first place."[19] Accordingly, race cannot be separated from economic or political power in attempting to explain the causes of disparate impact.

MORAL EVALUATION

Whether intentional or not, environmental policy in the United States is still perceived by movement members to be discriminatory and unjust, and thus, assailable on moral grounds. Witness the following statement by Chavis:

The environmental justice movement is a movement that confronts the immorality of upper- and middle-class people consuming the most energy and producing the most waste, while it is the health of the poor that is most affected by the resulting pollution.[20]

This emphasis on the health risks associated with exposure to environmental hazards is a frequent refrain among environmental justice advocates, strongly reflecting their moral condemnation of the perceived environmental inequities. Testifying before Congress, Chicago activist Hazel Johnson recited a list of health problems experienced by people in Altgeld Gardens, a public housing project encircled by so many polluting industries that it has been dubbed a "toxic doughnut." Noting the high incidence of cancer and other diseases among her family, friends, and neighbors, Johnson argues that the health conditions in her community are a "form of genocide."[21] Other activists talk about the "poisoning of their communities," which have become national "sacrifice zones."[22]

REMEDIAL ACTION

To remedy this situation, the environmental justice movement has demanded respect for the environmental rights of people of color. Laura Pulido[23] grouped these into two sets: (1) the right to participate in the regulatory process, and (2) the right to live free from pollution. This demand for both procedural and substantive rights was formally codified in the seventeen "Principles of Environmental Justice" adopted in Washington, D.C., at the First National People of Color Environmental Leadership Summit in October 1991. For example, principle seven "demands the right to participate as equal partners at every level of decision-making including needs assessment, planning, implementation, enforcement, and evaluation," while principle four asserts a "fundamental right to clean air, land, water, and food."[24]

As one activist put it: "The goal is equal justice and equal protection from pollution."[25] Within the environmental justice movement, however, differences of opinion exist concerning the relative emphasis that should be placed on strategies for achieving these rights. Some advocate a legislative and administrative approach, arguing that solutions must come from the federal level.[26] Others insist that the primary focus should be at the grassroots level, with community empowerment as the key to remedying environmental injustice.[27] Still others set their sights on nothing short of a structural transformation of the capitalist system, envisioning economic democracy as the only possibility for eliminating both the procedural and substantive violations of the rights of all citizens to a healthy environment.[28]

Environmental Justice as a Civil Rights Issue

Prior to the 1980s, there was little involvement of minority communities in environmental activism, for two main reasons. First, members of these communities believed that they could not afford the luxury of being primarily concerned about their environment when confronted by a plethora of pressing problems related to their day-to-day survival, such as poverty, unemployment, and inadequate housing. Second, the mainstream environmental movement at this time was overwhelmingly composed of white, middle- and upper-class members whose principal focus was preserving natural areas and endangered species. Thus, when African American neighborhood activists in the 1960s protested for improved garbage collection and sewer services, these were framed as "social" problems, not "environmental" ones.[29]

The construction of the environmental justice frame outlined above changed the perception of environmental issues in minority communities. Although the frame is not monolithic, it is possible to extract the following basic elements from the preceding discussion:

1. as a result of racial discrimination, intentional or otherwise, low-income communities of color are forced to bear a disproportionate share of environmental burdens;
2. these burdens are viewed as posing a serious health risk to community residents;
3. the solution to this problem lies in the pursuit of environmental justice; and
4. environmental justice can only be achieved by reducing pollution levels everywhere, and by granting full rights of democratic participation and self-determination to all communities threatened by environmental hazards.

By articulating its view of the problem in terms of racial discrimination and social justice, the environmental justice movement built on the rhetorical legacy of the civil rights movement. The collective action frame of the civil rights movement—which emphasized such values as individual rights, equal opportunities, social justice, full citizenship, human dignity, and self-determination—provided a "master frame" that legitimized the struggles of other disenfranchised groups. By framing the problem of disproportionate exposure as a violation of civil rights, the environmental justice movement was able to integrate environmental concerns into the civil rights frame.

This process of "frame extension" catalyzed the growth of the environmental justice movement. As Bullard put it:

> Blacks did not launch a frontal assault on environmental problems affecting their communities until these issues were couched in a civil rights context beginning in the early 1980s. They began to treat their struggle for environmental equity as a struggle against institutionalized racism and an extension of the quest for social justice. Just as black citizens fought for equal education, employment, and housing, they began to include the opportunity to live in a healthy environment as part of their basic rights.[30]

Other studies also attribute the resonance of the environmental justice frame to its connection with the civil rights frame.[31]

In addition to providing the environmental justice frame with a significant share of its rhetorical power, the civil rights movement has also contributed to the mobilizing potential of the environmental justice movement in two crucial respects. First, environmental justice activists have been able to draw on the organizational resources and institutional networks established during the previous struggle for racial equality. Churches, neighborhood improvement associations, and historically black colleges and universities have furnished the environmental justice movement with leadership, money, knowledge, communication networks, and other resources essential to the growth of any social movement. Second, environmental justice activists have successfully borrowed many of the tactics associated with the civil rights movement to call attention to their demands—both direct action tactics, such as protests and boycotts, as well as more conventional activities, such as lobbying and litigation.

This powerful combination of an interpretive frame, organizational resources, and mobilization techniques adapted from the civil rights movement is what has enabled the environmental justice movement to grow dramatically over the past fifteen years. As word of local struggles against toxic hazards spread, a number of statewide and regional networks of grassroots environmental justice organizations have sprung up across the country. In October 1991, these groups got together to convene the First National People of Color Environmental Leadership Summit.[32]

These statewide and regional networks have pressured both the mainstream environmental movement and the EPA to place environmental justice issues on their organizational agendas. Additional pressure on the EPA has come from academic researchers, who accelerated their study of the connection between race and the distribution of environmental burdens in the wake of the release of the 1987 Commission for Racial Justice report. In 1990, a group of activists and scholars met at the University of Michigan for a conference on "Race and the Incidence of Environmental Hazards."[33] The subsequent "Michigan Coalition" then lobbied EPA Administrator William Reilly to form a Work Group on Environmental Equity, which eventually led to the publication of an EPA report on environmental equity and the creation of the EPA's Office of Environmental Justice.[34]

As mentioned at the outset, in addition to the EPA, both Congress and the Clinton White House have recently elevated environmental justice issues to the formal policy agenda. In part, this is due to the political pressure generated by the movement, and to the mounting body of evidence suggesting that low-income and minority communities are subject to a disproportionate share of environmental hazards. Principally, however, the driving force behind the decision of policymakers to address the issue of environmental justice has been the ability of environmental justice activists to frame the problem as a civil rights issue.

The language of rights, the currency of much of our political discourse, is inherently linked to the two central ideas in our political culture: liberty and equality. As Richard Samp explains, one of the reasons for the tremendous success in recent times of the environmental justice movement has been its ability to put forward a pretty vague agenda. Certainly, almost everybody is for justice and against racism. In Samp's view, the concept of fairness is the strongest weapon in the movement's rhetorical arsenal: "The public has perceived the movement not as an effort to bring about a fundamental reorientation of society, but rather simply to distribute things in a fairer fashion."[35]

From Agenda Status to Policy Change?

Mobilizing activists and achieving agenda status is one thing; winning substantive policy change is another. In the second part of this chapter, I examine how other key participants in the policy process (Congress, the White House, and the EPA) have framed the issue of environmental justice, using the same categories employed above, and I discuss the various strategies proposed by the legislative and executive branches for achieving environmental justice. Although there is a general consensus regarding the definition of the problem, there remains considerable disagreement over the causes of disproportionate exposure to environmental hazards, the extent of the health risks associated with such exposure, and the appropriate remedies for this problem.

CONGRESS

Since 1992, a number of bills have been introduced in Congress, almost exclusively by Democrats, intended to address environmental justice and equity concerns. Additionally, several committees have held hearings on the subject. To date, none of the proposed legislation has been enacted, but an analysis of the Environmental Justice Act (EJA) of 1992 will prove instructive in revealing the differences between the environmental justice movement's framing of the issue and the legislative approach to the distribution of environmental hazards.

Introducing the EJA of 1992, then Senator Al Gore offered a similar definition of the problem to that endorsed by the environmental justice movement, expressing concern that our country

> faces disturbing inequities in the way severe pollutant problems are distributed. . . . In disproportionate amounts, toxic wastes and toxic emis-

sions from industrial processes contaminate the neighborhoods of minority communities.[36]

The introduction to the act itself declared that its goal was to establish a program to assure nondiscriminatory compliance with all environmental, health, and safety laws and to assure equal protection of the public health.

Nowhere in the proposed EJA, however, is there any suggestion that racial discrimination is responsible for the inequitable distribution of environmental hazards. Although one of the purposes of the act is to identify those areas with the highest levels of exposure to toxic chemicals, referred to as Environmental High Impact Areas (EHIAS),[37] there is no provision to identify the underlying causes of this disproportionate exposure.

Speaking publicly, members of Congress have tended to escalate the moral rhetoric regarding environmental justice. For example, Rep. McKinney (D–GA) declared that "the call to Environmental Justice is a visionary, community-centered call to let democracy prevail over the tyranny of callous polluters."[38] Yet the language of the EJA itself is dry and technical. As such, it fails to reflect the movement's perceptions of the urgent health risks facing certain minority communities. Moreover, the mathematical formulas used for calculating the existence of EHIAS do not even include any health-based criteria. As Linda Blank points out:

> It must be remembered that the proposed EJA is not just an environmental statute; it is also a civil rights statute. As such, it should specifically address human rights in clear, lucid language. By omitting health concerns from its initial analysis, EJA as currently written misses its very purpose and loses strength as a result.[39]

While the environmental justice movement demands both procedural and substantive solutions to the problem of disproportionate exposure, the proposed EJA offers little of the latter. The main provisions of the act call for the identification of EHIAS, to be followed by further studies, inspections, and reviews. Additionally, a technical assistance grant would be provided to one community group in each EHIA to facilitate public participation in the act's provisions. Blank is highly critical of the EJA for its lack of substantive elements:

Instead of providing needed solutions, EJA could just generate reports and studies and provide the foundation for drafting still more statutes. In the meantime, the environmental hazards disproportionately facing minority groups can only get worse.[40]

THE WHITE HOUSE

During the 1980s and early 1990s, the issue of environmental justice was ignored by the Reagan and Bush administrations. A major goal was to provide regulatory relief in the form of deregulation for the business sector. Undoubtedly, the most important shift in the political opportunity structure for the environmental justice movement was the election of Bill Clinton in 1992. The issue of environmental justice provided Clinton with an opportunity to direct a major policy initiative toward three key Democratic constituencies: the poor, minorities, and environmentalists. The commitment of the incoming administration to the cause of environmental justice was symbolized by the appointments of Chavis and Bullard to an environmental working group within the Clinton transition team.

The major contribution of the Clinton White House to the environmental justice policy debate has been Executive Order 12898. Here, too, the problem is defined in the same language as that employed by the environmental justice movement. The order instructs each federal agency to identify and address

> disproportionately high and adverse human health or environmental effects of its programs, policies, and activities on minority populations and low-income populations.[41]

As with the proposed EJA, there is no mention in Clinton's executive order of what is believed to be specifically responsible for the inequitable distribution of environmental hazards. Rather, the emphasis is on collecting sufficient data to make that determination at some future point. The order directs each federal agency to make environmental justice part of its mission, and instructs agencies to develop a strategy for identifying environmental justice concerns within their domain and for formulating appropriate responses. It also calls for the establishment of an interagency working group to provide guidance to federal agencies and to coordinate their efforts at implementing an environmental justice strategy.

As Gerald Torres points out, many people will dismiss the executive order as "just another process remedy." The order does not compel any particular substantive result, it only mandates consideration of environmental justice criteria in the regulatory process. The executive order, however, has the potential to transform the decision-making culture within administrative agencies, which will lead to a shift in the cognitive frame by which agencies approach the issue of environmental justice. As Torres explains:

Transforming the regulatory culture means transforming the basic analytical categories so that they incorporate things that they would not normally think would be part of the range of issues that they have to consider as they come to their substantive decision.[42]

THE ENVIRONMENTAL PROTECTION AGENCY

If the vision of environmental justice put forth by the movement is ever to be implemented, one federal agency that must transform its basic analytical categories is the EPA. In 1990, the EPA's Science Advisory Board issued a report endorsing "comparative risk analysis" as the basis for all environmental decision making.[43] This approach is based on a highly quantitative "scientific understanding of risk," as opposed to public risk perceptions, which were assumed to be "transported by middle-class enthusiasms."[44] Yet comparative risk analysis, by focusing on aggregate levels of risk, ignores questions of distributional equity and, thus, leaves little room for social justice criteria. A look at the EPA's 1992 Environmental Equity Report reveals how fundamentally different its interpretive frame is from that employed by the environmental justice movement. To begin with, as Eileen Gauna points out, "the difference in perspective is illuminated in the choice of terms used to define and address the problem of disparate environmental protection."[45] Witness the reasoning behind the EPA's decision to refer to the issue under study as one of "environmental equity":

EPA chose the term environmental equity because it most readily lends itself to scientific risk analysis. The distribution of environmental risks is often measurable and quantifiable. The Agency can act on inequities based on scientific data. Evaluating the existence of injustices and racism is more difficult because they take into account socioeconomic factors in addition to the distribution of environmental benefits that are beyond the scope of this report. Furthermore, environmental equity, in contrast to environmental racism, includes the disproportionate risk burden placed on any population group, as defined by gender, age, income, as well as race.[46]

Consequently, when the EPA report acknowledged that "racial minority and low-income populations experience higher than average exposures to selected air pollutants, hazardous waste facilities, contaminated fish, and agricultural pesticides," it labeled the problem as one of environmental inequity. In attempting to identify the causes of this disparate exposure, the possible role of intentional racial discrimination was downplayed, and poverty was asserted to be a more significant variable than race.[47] Commenting on the report, Robert Wolcott, the chair of the Workgroup, asked rhetorically:

> Is there systematic racism out there? I don't think so. It's more economic class. It comes down to resources to locate oneself in jobs and homes that avoid exposure. In many cases, racial minorities don't have the capital to exercise that mobility. . . . It doesn't seem to be the result of any venal intent. Real estate markets play themselves out.[48]

Ironically, although Wolcott does implicitly acknowledge the relationship between race and class, his framing of the causal forces behind the problem of disproportionate impact emphatically denies any connection between the two variables.

Environmental justice activists were highly critical of the EPA's analysis of the causal factors underlying the distribution of environmental risks. In their comments appended to the report, they took the agency to task for ignoring the existence of institutionalized forms of racial discrimination—such as housing discrimination, land use planning, and redlining—in explaining environmental inequities. In addition, members of the Michigan Coalition chastised the EPA for neglecting to include in its scientific literature review a range of studies demonstrating the significance of race as a key variable in accounting for the distribution of environmental hazards.[49]

The report's recommendations generally call for greater attention to equity issues, primarily through more data collection and improvements in the processes of "risk assessment" and "risk communication." The general tone of the report is communicated in technical language, which implicitly suggests that these processes are value neutral. This stands in marked contrast to the environmental justice movement frame, with its explicit moral condemnation of institutionalized racist practices and its urgent characterization of the health risks facing minority and low-income communities exposed to disproportionate levels of toxic hazards. When the EPA report asserted that there was a lack of sufficient data to determine the level of adverse health effects associated with this exposure, environmental justice activists responded angrily. Richard Moore, co-chair of the Southwest Network for Environmental and Economic Justice, commented:

> We want protection, not another study. . . . We've studied this issue to death. When you see poor communities that have six times more miscarriages than they should have or clusters of babies born without brains, you don't need another study to tell you something is wrong.[50]

Frame Differences: Social Justice versus "Solid Science"

Moore's remark illustrates perhaps the fundamental difference between the environmental justice movement frame and that advanced by the other key participants in the policy process. Movement members argue that the major statistical studies undertaken to date conclusively demonstrate that polluting facilities are disproportionately located in minority and low-income communities. Consequently, they contend that residents of these communities are disproportionately exposed to the health hazards these facilities pose, suffering severe health impacts as a result. Moreover, activists demand specific and immediate responses to environmental injustices. Conversely, government officials—joined by industry representatives—respond by ignoring or denying intentional racial discrimination, and by calling for more investigation and study of the health risks associated with exposure to toxic hazards. Activists, in turn, charge that governmental decisions to study the problem further is a delaying tactic intended to provide only symbolic recognition of the problem.[51] And, indeed, the credibility of the EPA's concern for the problems of environmental inequity was severely undermined when an internal memo was leaked the same month as the agency's 1992 Environmental Equity Report was released. In the memo, an associate administrator advocates a two-track strategy directed at both mainstream and grassroots environmental organizations. It is worth quoting at some length:

> The goal of this strategy is to win the recognition the agency deserves for its environmental equity and cultural diversity programs before the minority fairness issue reaches the "flashpoint"—that stage in an emotionally charged public controversy when activist groups finally succeed in persuading the more influential mainstream groups (civil rights organizations, unions, churches) to take ill-advised actions. From what we've begun seeing in the news, this issue is reaching that point.
>
> . . . We may be able to deflect some of the [activist groups'] hostility by taking the initiative to grant them respect and the access they want and by finding some common ground on which we can agree. Talking with and listening to these groups is the second track of the strategy.[52]

Given the range of variables associated with both exposure to toxic facilities and the health risks that they potentially pose, it is difficult to dispute the assertion that additional data regarding the causes and consequences of environmental inequity would be beneficial in determining appropriate remedies. The question arises, however, as to whether or not conventional science is equipped to provide the types of answers that would address the concerns raised by the environmental justice movement. As observed above, comparative risk analysis, the type of science practiced by the EPA, is ill equipped to evaluate environmental policymaking from a civil rights perspective. Moreover, the scientific emphasis on demonstrating certainty, or at least statistically significant causal relationships, presents the image of science as an objective, value-free process, which it clearly is not.[53] As Sylvia Tesh argues, "to call for absolute certainty and agreement among scientists before taking preventive action is merely a delaying tactic, effective only to the extent that people believe the myth that certainty characterizes science."[54]

Environmental Justice: Procedural versus Substantive Remedies

The solutions incorporated into the environmental justice movement frame demand both procedural rights (e.g., access to information, fair hearings, meaningful participation in the decision-making process) and substantive ones (e.g., reduction of toxic threats to all communities, increased employment opportunities, better housing, improved health care). As the above survey of official responses to movement claims demonstrated, government is more likely to offer the former. Although political scientists have long noted this phenomenon, some argue that a real possibility exists for procedural responses to lead to substantive policy changes, if movements are capable of "insinuating themselves into the established working relations between government and business."[55]

To date, the procedural remedies offered by the Clinton administration do not appear to offer much in the way of substantive changes in environmental policy. Despite EPA Administrator Browner's claim that her agency is committed to giving people access to the decision-making process about what happens in their communities and neighborhoods, the performance of the Interagency Working Group on Environmental Justice, established by Executive Order 12898, has demonstrated otherwise. Just three weeks prior to the deadline for the submission of its proposed environmental justice strategy, the working group held its first public meeting.

While environmental justice activists were not expecting the executive order to be a "silver bullet," they did hope the process could serve as an opening for activists to work directly with federal agencies in shaping their policies. Instead, the results were fraught with political hype and empty of real solutions. Panelists were unprepared for engaging in substantive dialogue, offering little more than platitudes to activists demanding answers to serious problems. Moreover, when the working group's recommendations were finally presented a few weeks later, they were acknowledged to be more "consciousness-raising than prescriptive," focusing on "options" as opposed to "action plans."[56]

Legal Remedies

Given the decentralized nature of political power in America, social movements are often compelled to press their claims in a variety of institutional and noninstitutional settings. When elected officials fail to support a movement's agenda, some movement organizations turn to the courts to redress their grievances. For over 150 years, groups struggling for recognition of their full rights as citizens have been able to draw on the constitutional doctrines of equal protection and due process enunciated in the Fourteenth Amendment.

It is hardly surprising, then, given the civil rights orientation of the environmental justice movement's collective action frame, as well as the movement's ties to civil rights organizations, that environmental justice claims would be pursued through legal channels. In 1979, a group of neighborhood activists in Houston, Texas, filed the first of several environmental justice lawsuits alleging a violation of civil rights on equal protection grounds. The lawsuit challenged the siting of a hazardous waste facility in a predominantly African American neighborhood and attempted, through statistical analysis, to demonstrate a pattern of racially discriminatory siting decisions.[57]

This effort, and others like it,[58] proved to be unsuccessful. Under current Supreme Court doctrine, plaintiffs in equal protection cases must conclusively demonstrate discriminatory intent.[59] This standard is a nearly insurmountable obstacle when trying to prove environmental discrimination, even when the Court agrees that the result will produce a disproportionate impact on minority communities.

Given the difficulty of demonstrating environmental racism using a constitutional argument, some legal scholars and activists have called for alternative approaches to environmental justice litigation based on existing statutory law.[60] Many environmental statutes, including the National Environmental Policy Act (NEPA), offer procedural mechanisms for challenging official decisions.[61] In Kettleman City, California, for example, a group of Latino farmworkers prevented the permitting of a hazardous waste incinerator on the grounds that the failure to translate the mandated environmental impact report into Spanish constituted a violation of the public participation requirement of the California Environmental Quality Act, which was modeled on the NEPA.[62]

Not content to cede the utility of a civil rights approach to environmental justice litigation, some activists have turned to Title VI of the 1964 Civil Rights Act, which prohibits discrimination on the basis of race, color, or national origin in all programs or activities that receive federal funding. One advantage of this approach is that the EPA's standard for deciding Title VI claims is one of discriminatory impact or effects, not intent. Thus, under Title VI, federal funds can be denied to states that take part in environmental decisions that can be shown to have a discriminatory effect.

In a memorandum accompanying Executive Order 12898, President Clinton directed the EPA and all other federal agencies to enforce Title VI with respect to environmental policies.[63] Previously, the EPA held that Title VI regulations did not apply to the agency, since its mission was viewed as a scientific and technical one; social concerns were not part of the EPA's mandate. In March 1993, however, the EPA announced a change in its position, acknowledging that Title VI regulations do fall within its purview.[64] As a result, the EPA's Office of Civil Rights has undertaken a number of Title VI investigations, at least one of which has led to the cancellation of a hazardous waste permit in a predominantly African American community in Louisiana.

Despite the promise offered by a Title VI approach to remedying distributional inequities in the siting of toxic facilities, the environmental justice movement must adopt a cautious stance toward the use of legal tactics. In the past five years, the legal community has displayed considerable interest in the issue of environmental justice: filing lawsuits, offering law school courses, establishing environmental justice legal clinics, and publishing dozens of law review articles. Yet there is a danger to this remedial action to the extent that legal strategies shift movement struggles out of the control of local activists and into the hands of lawyers and national legal organizations. As a result, laws and litigation are seen as the solution to the problem of environmental injustices, as opposed to community empowerment. Instead of being viewed as a means of achieving fundamental systemic change, "environmental justice" becomes redefined and transformed into just another issue for which we need legislation, or a new legal strategy. The entrance of legal groups into the environmental justice field is, in many ways, a detriment to the movement, blunting its ideological edge and diverting its limited resources.

Conclusion

This suggests that the legal community, in its approach to environmental justice issues, employs a different frame than the one adopted by grassroots activists. At one level, this can be seen in the "discriminatory intent" hurdle placed by the courts in the path of those pursuing an equal protection claim, or the tendency of legal organizations to advocate legal solutions, even at the expense of community empowerment. At a more fundamental level, it means that adopting a legal strategy for remedying environmental inequities requires talking about the problem in complex legal terms, just as pursuing an EPA-oriented strategy requires talking about the problem in technical, scientific language. Neither level represents the type of discourse that mobilized the environmental justice movement in the first place.

By incorporating environmental concerns into a civil rights frame, the environmental justice movement was able to characterize the distribution of environmental hazards as part of a broader pattern of social injustice, one that contradicted fundamental beliefs of fairness and equality. By articulating its grievances from this perspective, the movement has grown substantially in a relatively short period of time, generating sufficient political pressure to secure the attention of government officials. While there is a role for both scientific analysis and legal tactics in the quest for environmental justice, political pressure is likely to subside if the movement engages in a debate that takes place within a scientific or legal frame, as opposed to a political one.

As government officials, industry representatives, and academics engage in an ongoing debate over the causes of the inequitable distribution of environmental hazards and the most appropriate solutions to this problem, the movement must continue to point to larger questions about the distribution of political and economic power in society, and the values that should inform environmental policymaking. In doing so, environmental justice activists must maintain their rhetorical emphasis on civil rights and their grassroots organizing strategy, for these are clearly the movement's strengths. At the same time, they must continue to demand immediate action to remedy the problems associated with disproportionate exposure to toxic threats. As one commentator observed:

> The way people . . . want to argue about this issue is that they want to tell you what causes it and they want to tell you what labels to put on it because they think that points to a solution. . . . My own view is that the causes of the fact that black children live in a ghetto and have blood levels

[of lead] twice as high as rich white kids are irrelevant. The key thing is that the government—the society—has a responsibility to put an end to that type of situation. If a debate about putting an end to it starts with putting a label on it or assigning a cause to it, as far as I'm concerned, that gets in the way.[65]

Notes

1. Robert D. Bullard, *Dumping in Dixie: Race, Class, and Environmental Quality,* 2d ed. (Boulder, Colo.: Westview Press, 1994), 29–31.

2. Robert D. Bullard, *People of Color Environmental Groups Directory* (Flint, Mich.: Charles Stewart Mott Foundation, 1992).

3. This chapter will not examine state-level approaches to environmental justice, but a considerable amount of activity is taking place in several states, particularly in the South, which could have a significant impact on both local struggles and national policymaking. For an overview, see Stacy Hart, "A Survey of Environmental Justice Legislation in the States," *Washington University Law Quarterly* 73 (1995): 1459–75.

4. See Peter Berger and Thomas Luckmann, *The Social Construction of Reality* (Garden City, N.Y.: Doubleday, 1967); Joel Best, ed., *Images of Issues: Typifying Contemporary Social Problems* (New York: Aldine de Gruyter, 1989); Herbert Blumer, "Social Problems as Collective Behavior," *Social Problems* 18 (1970): 298–306; James Holstein and Gale Miller, eds., *Reconsidering Social Constructionism: Debates in Social Problems Theory* (New York: Aldine de Gruyter, 1993); and Malcolm Spector and John Kitsuse, *Constructing Social Problems* (Menlo Park, Calif.: Cummings, 1977).

5. The literature on social movements is extensive. For a useful overview of the resource mobilization approach, two helpful sources are Doug McAdam, John McCarthy, and Mayer Zald, "Social Movements," in *Handbook of Sociology,* ed. Neil Smelser (Newbury Park, Calif.: Sage, 1988); and Mayer Zald, "Looking Backward to Look Forward: Reflections on the Past and Future of the Resource Mobilization Research Program," in *Frontiers in Social Movement Theory,* ed. Aldon Morris and Carol McClurg Mueller (New Haven, Conn.: Yale University Press, 1992), 326–48.

6. See Morris and Mueller, *Frontiers in Social Movement Theory,* chaps. 1–8.

7. Robert M. Entman, "Framing: Toward Clarification of a Fractured Paradigm," *Journal of Communication* 43, no. 4 (fall 1993): 52.

8. David A. Snow and Robert D. Benford, "Master Frames and Cycles of Protest," in Morris and Mueller, 137.

9. Rhys Williams, "Constructing the Public Good: Social Movements and Cultural Resources," *Social Problems* 42, no. 1 (February 1995): 126.

10. Ibid.

11. E. E. Schattschneider, *The Semisovereign People* (New York: Holt, Rhinehart, and Winston, 1960), 66.

12. On the relationship between social construction and agenda setting, see Roger W. Cobb and Charles Elder, *Participation in American Politics: The Dynamics of Agenda Building*, 2d ed. (Baltimore, Md.: Johns Hopkins University Press, 1983); Murray Edelman, *Political Language: Words That Succeed and Policies That Fail* (New York: Academic Press, 1977); John Kingdon, *Agendas, Alternatives, and Public Policies* (Boston: Little, Brown, 1984); and David A. Rochefort and Roger W. Cobb, eds., *The Politics of Problem Definition: Shaping the Policy Agenda* (Lawrence: University Press of Kansas, 1994).

13. Snow and Benford, "Master Frames and Cycles of Protest," 137.

14. Robert D. Bullard, "Overcoming Racism in Environmental Decision-Making," *Environment* (May 1994): 11.

15. Paul Mohai and Bunyan Bryant, "Environmental Racism: Reviewing the Evidence," in *Race and the Incidence of Environmental Hazards: A Time for Discourse*, ed. Bunyan Bryant and Paul Mohai (Boulder, Colo.: Westview Press, 1992). The sixteen studies cover a period from 1971 to 1992. The quote is taken from Paul Mohai and Bunyan Bryant, "Environmental Injustice: Weighing Race and Class as Factors in the Distribution of Environmental Hazards," *University of Colorado Law Review* 63 (1992): 927.

16. Marianne Lavelle and Marcia Coyle, "Unequal Protection: The Racial Divide in Environmental Law," *National Law Journal* (21 September 1992): S2–S12.

17. Testimony of Dr. Benjamin Chavis, House Subcommittee on Civil and Constitutional Rights, *Hearings on Environmental Justice 1993*, 6.

18. Testimony of Robert D. Bullard, House Subcommittee on Civil and Constitutional Rights, *Hearings on Environmental Justice 1993*, 48.

19. Robert R. Higgins, "Race and Environmental Equity: An Overview of the Environmental Justice Issue in the Policy Process," *Polity* 26, no. 2 (winter 1993): 287.

20. Benjamin Chavis, "Environmental Justice is Social Justice," *Los Angeles Times*, 19 January 1993, B7.

21. Testimony of Hazel Johnson, House Subcommittee on Civil and Constitutional Rights, *Hearings on Environmental Justice 1993*, 8.

22. Testimony of Pat Bryant, House Subcommittee on Civil and Constitutional Rights, *Hearings on Environmental Justice 1993*, 9–10.

23. Laura Pulido, "Restructuring and the Contraction and Expansion of Environmental Rights in the United States," *Environment and Planning A* 26 (1994): 919.

24. The "Principles of Environmental Justice" are reprinted in Richard Hofrichter, ed., *Toxic Struggles: The Theory and Practice of Environmental Justice* (Philadelphia: New Society Publishers, 1993), 237–39.

25. Robert D. Bullard, ed., *Unequal Protection: Environmental Justice and Communities of Color* (San Francisco: Sierra Club Books, 1994), 316.

26. See Deeohn Ferris, "A Broad Environmental Justice Agenda: Mandating Change Begins at the Federal Level," *Maryland Journal of Contemporary Legal Issues* 5, no. 1 (1995): 115–27; and Gerald Torres, "Changing the Way Government Views Environmental Justice," *St. John's Journal of Legal Commentary* 9 (1994): 543–54.

27. See Robert D. Bullard, ed., *Confronting Environmental Racism: Voices from the Grassroots* (Boston: South End Press, 1993).

28. See John O'Connor, "The Promise of Environmental Democracy," in Hofrichter, *Toxic Struggles*, 47–57; and Anthony Taibi, "Environmental Justice, Structural Economic Theory, and Community Economic Empowerment," *St. John's Journal of Legal Commentary* 9 (1994): 481–507.

29. Bullard, *Confronting Environmental Racism*, 9. This is not to suggest that, prior to the 1980s, African Americans had not demonstrated any concern for environmental issues. See Paul Mohai, "Black Environmentalism," *Social Science Quarterly* 71 (1990): 744–65.

30. Bullard, *Dumping in Dixie*, 29.

31. For other versions of the environmental justice frame, and its connection with the civil rights movement, see Stella Capek, "The 'Environmental Justice' Frame: A Conceptual Discussion and an Application," *Social Problems* 40, no. 1 (1993): 5–21; and Debra Salazar and Lisa Moulds, "Toward an Integrated Politics of Social Justice and the Environment: African American Leaders in Seattle" (paper presented at the annual meeting of the Western Political Science Association, Portland, Oreg., March 16–18, 1995). See also Bullard, *Dumping in Dixie*, 118–22.

32. The regional organizations included the Southwest Network for Environmental and Economic Justice, the Southwest Organizing Project, the Southern Organizing Committee, the Gulf Coast Tenants Organization, the Indigenous Environmental Network, and others. For more on the 1991 national conference, see United Church of Christ Commission for Racial Justice, *Proceedings of the First National People of Color*

Environmental Leadership Summit (New York: United Church of Christ, 1991); and Karl Grossman, "The People of Color Environmental Summit," in Bullard, *Unequal Protection,* 272–77.

33. See Bunyan Bryant and Paul Mohai, eds., *The Proceedings of the Michigan Conference on Race and the Incidence of Environmental Hazards* (Ann Arbor: School of Natural Resources, University of Michigan, 1990); and Bryant and Mohai, *Race and the Incidence of Environmental Hazards.*

34. See U.S. Environmental Protection Agency, Office of Policy, Planning, and Evaluation, *Environmental Equity: Reducing Risk for All Communities,* 2 vols., EPA–230–R–92–008 (Washington, D.C.: Environmental Protection Agency, June 1990). Vol. 1 is titled the *Workgroup Report to the Administrator* and Vol. 2 is titled the *Supporting Document.*

35. Richard Samp, "Fairness for Sale in the Marketplace," *St. John's Journal of Legal Commentary* 9 (1994): 503.

36. Cited in Claire Hasler, "The Proposed Environmental Justice Act: I Have a (Green) Dream," *University of Puget Sound Law Review* 17, no. 2 (winter 1994): 417–18.

37. H.R. 2105, 103rd Cong., 1st. sess. (1993): Title I.

38. Statement of Rep. Cynthia McKinney, House Subcommittee on Civil and Constitutional Rights, *Hearings on Environmental Justice 1993,* 184.

39. Linda D. Blank, "Seeking Solutions to Environmental Inequity: The Environmental Justice Act," *Environmental Law* 24 (1994): 1123.

40. Ibid., 1121.

41. Executive Order 12898, section 1–101. President Clinton, "Federal Action to Address Environmental Justice in Minority Populations and Low Income Populations," 11 February 1994.

42. Torres, "Changing the Way," p. 550.

43. U.S. Environmental Protection Agency, Science Advisory Board, Relative Risk Reduction Strategies Committee, *Reducing Risk: Setting Priorities and Strategies for Environmental Protection* (Washington, D.C.: Environmental Protection Agency, 1990).

44. Statement of William Reilly, Senate Committee on Environment and Public Works, *Reducing Risk, Setting Priorities and Strategies for Environmental Protection: Hearings on Recent Science Advisory Board Report,* 102nd Cong., 1st sess. (1991): 48. For an analysis of how the EPA could better respond to public perceptions of risk, see James Freeman and Rachel D. Godsil, "The Question of Risk: Incorporating Community Perceptions into Environmental Risk Assessments," *Fordham Urban Law Journal* 21, no. 3 (spring 1994): 547–76.

45. Eileen Gauna, "Federal Environmental Citizen Provisions: Obstacles and Incentives on the Road to Environmental Justice," *Ecology Law Quarterly* 22, no. 1 (1995): 26 n. 86.

46. EPA, *Environmental Equity,* vol. 2, 2–3.

47. EPA, *Environmental Equity,* vol. 1, 3.

48. Cited in Michael Weisskopf, "Minorities' Pollution Risk Is Debated; Some Activists Link Exposure to Racism," *Washington Post,* 16 January 1992, A25.

49. EPA, *Environmental Equity,* vol. 2, 78.

50. Cited in Gauna, "Federal Environmental Citizen Provisions," 17 n. 54.

51. See Murray Edelman, *The Symbolic Uses of Politics* (Urbana: University of Illinois Press, 1964).

52. See Lewis Crampton, "Environmental Equity Community Plan," confidential EPA memorandum to Gordon Bender, chief of staff, 15 November 1991, cited in Bullard, *Confronting Environmental Racism.*

53. See Bunyan Bryant, "Issues and Potential Policies and Solutions for Environmental Justice: An Overview," in Bryant, ed., *Environmental Justice: Issues, Policies, and Solutions* (Washington, D.C.: Island Press, 1995), 8–23.

54. Sylvia Tesh, *Hidden Arguments: Political Ideology and Disease Prevention Policy* (New Brunswick, N.J.: Rutgers University Press, 1990), 69, cited in Bryant, "Issues and Potential Policies," 10.

55. See Thomas Rochon and Daniel Mazmanian, "Social Movements and the Policy Process," *Annals of the American Academy of Political and Social Science* 528 (July 1993): 75–87.

56. Jeff Johnson, "Federal Environmental Justice Plans Go to Clinton," *Environmental Science and Technology* 29, no. 4 (1995): 22.

57. The case was *Bean v Southwestern Waste Management Corporation,* 482 F Supp 673 (SD Tex 1979), affirmed without opinion, 782 F2d 1038 (5th Cir 1986). The article cited in the previous note contains a discussion of this case.

58. See also *East Bibb Twiggs Neighborhood Association v Macon-Bibb County Planning and Zoning Commission,* 706 F Supp 880 (MD Ga), affirmed, 896 F2d 1264 (11th Cir 1989); *RISE, Inc. v Kay,* 768 F Supp 1141 (ED Va 1991); *RISE, Inc. v Kay,* 768 F Supp 1144 (ED Va 1991); and *Twitty v State,* 354 SE2d 296 (NC Ct App 1987).

59. See *Village of Arlington Heights v Metropolitan Housing Development Corporation,* 429 US 252 (1977); and *Washington v Davis,* 426 US 229 (1976).

60. See Alice L. Brown, "Environmental Justice: New Civil Rights

Frontier," *TRIAL* (July 1993): pp. 48–53; and Luke Cole, "Environmental Justice Litigation: Another Stone in David's Sling," *Fordham Urban Law Journal* 21, no. 3 (spring 1994): 523–45.

61. See Heather Ross, "Using NEPA in the Fight for Environmental Justice," *William and Mary Journal of Environmental Law* 18 (1994): 353–74; and Melany Earnhardt, "Using the National Environmental Policy Act to Address Environmental Justice Issues," *Clearinghouse Review* 29, no. 4 (1995): 436–45.

62. For a detailed discussion of the Kettleman City case, see Cole, "Environmental Justice Litigation."

63. President William J. Clinton, "Memorandum on Environmental Justice" (11 February 1994). Public Papers of the President (Washington, D.C.: Government Printing Office), 241–42.

64. See testimony of Dr. Clarice Gaylord, House Subcommittee on Civil and Constitutional Rights, *Hearings on Environmental Justice Hearings.*

65. George Van Cleve, "Equal Enforcement for All," *St. John's Journal of Legal Commentary* 9 (1994): 527–28.

II

Environmental Injustices

Harvey L. White

RACE, CLASS, AND ENVIRONMENTAL HAZARDS

When we talk about environmental justice, we mean calling a halt to the poisoning of our poorest communities, from our rural areas to our inner cities. We don't have a person to waste and pollution clearly wastes human lives and natural resources. When our children's lives are no longer damaged by lead poisoning, we will stop wasting the energy and intelligence that could build a stronger and more prosperous America.
—President Bill Clinton, June 14, 1993

As these remarks by President Clinton suggest, environmental hazards represent major health concerns for urban and rural communities. Arguably, they are one of the greatest health risks facing this country for all Americans—Euro-Americans, people of color, rich and poor, old and young. In several respects, environmental hazards pose a greater health threat than the dreaded AIDS virus.

In relation to AIDS, individuals can take steps to protect themselves. They can either restrict their exposure to the virus by abstaining from risky sexual and drug-related activities or other activities that constitute mediums through which the virus is transmitted. In other words, they can isolate themselves almost completely from transmittal mediums. Or, if they choose to expose themselves to these mediums, there are protective measures readily available that can be employed to minimize the risk of contracting AIDS.

In contrast, when environmental hazards exist there is virtually nothing that individuals can do to protect themselves. Most crucial, they cannot limit their exposure to the mediums that transmit environmental hazards, which are the air, food, and water necessary for survival. Furthermore, there are no measures that are readily available, which can easily be employed, to minimize the risk of coming in contact with environmental hazards.

Table 1. Selected Studies of Racial and Income Disparities in the Distribution of Environmental Hazards, 1967–1993

Year	Author	Type of Hazards	Geographic Focus	Disparity Race	Disparity Income
1967	Hoffman et al.	Pesticides	Chicago, Ill.	Yes	
1971	CEO	Air pollution	Chicago, Ill.		Yes
1972	Davis et al.	Pesticides, blood level	Dade County, Fla.		Yes
1972	Freeman	Air pollution	Kansas City/St. Louis/D.C.	Yes	Yes
1974	Burns	Pesticides	Southern states	Yes	
1975	Kruvant	Air pollution	Washington, D.C.	Yes	Yes
1975	Zupan	Air pollution	New York, N.Y.	Yes	
1976	Bruch	Air pollution	New Haven, Conn.	No	Yes
1977	Berry et al.	Pollution/pesticides, etc.	Urban areas	Yes	Yes
1977	Kutz et al.	Pesticides	National	Yes	
1978	Asch and Seneca	Air pollution	Urban areas	Yes	Yes
1980	SRI	Toxic fish	National	Yes	No
1981	Puffer	Toxic fish	Los Angeles, Calif.	Yes	
1983	U.S. GAO	Hazardous waste	Southeast	Yes	
1984	Greenberg and Anderson	Hazardous waste	New Jersey	Yes	Yes
1985	McAllum	Toxic fish	Puget Sound, Wash.	Yes	
1985	NOAA	Toxic fish	Puget Sound, Wash.	Yes	
1986	Gould	Hazardous waste	National		Yes
1987	UCC and PDA	Hazardous waste	National	Yes	Yes
1987	Gelobter	Air pollution	Urban areas	Yes	Yes
1988	ATSDR	Lead	Urban areas	Yes	Yes
1989	Belliveau et al.	Toxic releases	Richmond, Calif.	Yes	Yes
1989	Pfaff	Air pollution	Detroit, Mich.		Yes
1990	Cater-pokras et al.	Lead	National		Yes
1991	Brown	Toxic releases	St. Louis, Mo.	Yes	
1991	Costner and Thornton	Hazardous waste	National	Yes	Yes
1991	Kay	Toxic releases	Los Angeles, Calif.	Yes	
1991	Mann	Air pollution	Los Angeles, Calif.	Yes	
1991	Wernette and Nieves	Air pollution	Urban areas	Yes	
1992	Fitton	Hazardous waste	National	Yes	Yes
1992	Goldman	Toxic air/waste	National	Yes	No
1992	Holtzman	Waste incineration	New York, N.Y.	Yes	
1992	Ketkar	Hazardous waste	New Jersey	Yes	
1992	McDermott	Hazardous waste	National	Yes	
1992	Mohai and Bryant	Hazardous waste	Detroit, Mich.	Yes	Yes
1992	Nieves	Toxic waste/pollution	National	Yes	Yes
1992	Roberts	Hazardous waste	New York, N.Y.		Yes
1992	Unger et al.	Hazardous waste	Pinewood, S.C.	Yes	Yes
1992	West et al.	Toxic fish	Michigan	Yes	No
1993	Been	Hazardous waste siting	Southeast	Yes	Yes
		Postsiting of hazards	Southeast	Yes	Yes
1993	Burke	Toxic releases	Los Angeles, Calif.	Yes	Yes
1993	Bowen et al.	Toxic releases	Cuyahoga, Ohio	No	Yes
		Toxic releases	Ohio	Yes	No
1993	Greenberg	Incinerators (large)	National	Yes	No
1993	Hamilton	Hazardous waste siting	National	Yes	Yes
1993	Zimmerman	Hazardous waste	National	Yes	No

Source: Derived from Benjamin A. Goldman, *Not Just Prosperity: Achieving Sustainability with Environmental Justice* (Washington, D.C.: National Wildlife Federation, 1993).

Once environmental hazards are in the mediums that transmit them, the very life-sustaining functions that individuals must perform may put them at risk of exposure to life-threatening toxins. Thus, it is almost impossible to protect oneself from environmental hazards because individuals have virtually no control over the quality of the air they breathe, the food they eat, or the water they drink. They are almost completely dependent on someone else to protect them from environmental hazards. Consequently, every individual experiences risk from the toxic pollution that threatens our planet.

Yet not all individuals are equally at risk. Studies of environmental hazards indicate that there are significant racial and economic disparities in the distribution of risks (see table 1). Racial disparities were found in 87 percent of the studies and income disparities were found in 74 percent. Disparities were found to exist in a variety of areas (i.e., exposure to toxins and solid waste, siting of hazardous facilities, and occupational health), all regions of the country, and in both urban and rural communities.[1] In other words, if you are African American, Native American, Latino, or poor, you are likely to be at risk from environmental hazards more frequently.

Research also suggests that people of color and the poor are often more severely exposed to potentially deadly and destructive levels of toxins from environmental hazards than others. The nature of the endangerment experienced by some of these individuals is life-threatening. For others, particularly the young, it can be debilitating. Findings from this research constitute convincing evidence that this pattern of exposure transcends almost every aspect of their lives: the places where they *work, live, play,* and *learn;* and in the *foods* they eat.

The scholars who conducted the studies listed in table 1 are from thirteen different professional fields and used a variety of research methods to complete their work. Their findings are more than just a persuasive body of evidence; they also suggest the need for concern about environmental injustice in the United States. The following is an overview of data from these and other studies.

Racial and Socioeconomic Disparities in the Workplace

Whatever the health effects are of environmental hazards on those who live near facilities that generate them, the impact on workers is likely to be more severe. Workplace exposure is generally more direct, continual, and concentrated.[2] It is estimated that as many as 50,000 to 70,000 workers in the United States die from occupational diseases annually, and new cases of work-related illnesses are believed to be between 125,000 and 350,000 each year.[3] The EPA has concluded that workplace exposure to environmental hazards poses a greater health risk than any other known factor. As in other instances, however, this risk is not evenly distributed.[4]

It is no secret that the poor and people of color are usually hired for the worst jobs. These hiring practices have many consequences. Exposure to environmental hazards is just one, albeit an important one. Individuals in these racial and economic groups often occupy the most strenuous and hazardous jobs. Such jobs are also likely to be those that pose the greatest risk of exposure to chemicals and substances that can be detrimental to one's health.

African Americans and other people of color, in particular, have been found to bear a disproportionate share of the occupational risks emanating from environmental hazards in the workplace. For instance, researchers have learned that African Americans have a 37 percent greater chance of suffering an occupationally induced injury or illness, and a 20 percent greater chance of dying from an occupational disease or injury, than do white workers. Black workers are almost twice as likely to be partially disabled because of job-related injuries or illnesses.[5]

Studies of industries where large numbers of African American workers are employed reveal a significantly disproportionate exposure to cancer-causing substances (African American workers in these industries also have elevated levels of several types of cancer). A study of 6,500 rubber workers, in a tire manufacturing plant in Akron, Ohio, found that 27 percent of African American workers had been exposed to dust, chemicals, and vapor particles that contained toxins; only 3 percent of the white workers experienced similar exposure.[6]

In a study of 59,000 steel workers, it was revealed that 89 percent of nonwhite coke plant employees had been assigned to the coke oven area (one of the most hazardous aspects of steel production), while only 32 percent of white employees had worked in that area of the plant. Nonwhite employees in the coke plant experienced double the expected cancer-related death rate.[7]

A U.S. Public Health Department study of chromate workers found that the expected cancer mortality rate for African Americans was an alarming 80 percent; it was 14.29 percent for whites.[8] Similar findings were discerned in a cancer mortality study of coastal Georgia residents. This study discovered that African American shipyard workers had a lung cancer death rate two times higher than expected.[9]

The pattern of industrial exposure described above has been observed in the agricultural sector as well, where an estimated 313,000 farmworkers in the United States may suffer from pesticide-related illnesses each year.[10] Ivette Perfecto calculates that 90 percent of the approximately two million U.S. farmworkers are people of color.[11] For a great many years, researchers have found that most farm pesticide exposures occur among low-income Latino and African American migrant workers.[12] Agriculture has become the third most dangerous occupation in the United States. According to the National Safety Council, the death rate in agriculture is 66 per 10,000, while the industrial average is only 18 per 10,000.[13] Workers in this sector, who are mostly low-income individuals of color, have some of the most dangerous and least-protected jobs.

The findings described above and those from other studies led the National Institute for Occupational Safety and Health (NIOSH) to conclude that "minority workers tend to encounter a disproportionately greater number of serious safety hazards because they are employed in especially dirty and dangerous jobs."[14] NIOSH's conclusion is supported by data indicating that mortality from acutely hazardous work exposure among men of color is 50 percent higher than it is among white men.[15] In addition to the workplace, people of color, particularly those who are economically distressed, are also exposed more frequently and severely to environmental hazards where they live, learn, and play.

How Safe Is It Where We Live, Learn, and Play?

Many studies have examined the siting of facilities that are considered to be environmental hazards.[16] Although economics, or the class variable, was discerned to be an important factor, the conclusion drawn in these and most other studies is that environmental hazards are located disproportionately in communities where people of color and the poor live, learn, and play.

In counties that rank the worst across all industrial toxin measures, people of color comprise more than twice the percentage of the population than is the average for the rest of the country. For example, the largest hazardous waste landfill, which receives toxic materials from forty-five states and several foreign countries, is located in predominantly African American and poor Sumpter County in the heart of the Alabama Black Belt.[17] In Houston, Texas, six out of eight municipal incinerators are located in principally African American neighborhoods. One of the other two is located in a mainly Latino neighborhood.[18]

Cancer-causing asbestos, found to be prevalent in Chicago housing projects, is believed to be a serious problem common to most of the nation's inner-city housing, where a large percent of the residents are people of color and poor.[19] Particularly disturbing is emerging evidence suggesting that children of color from low-income families often experience more severe and frequent exposures to environmental hazards than adults.

Recent reports indicate that children of color who are poor not only are more likely to live in homes with peeling lead paint,[20] they are also more likely to play in parks that are contaminated and attend schools that contain toxins.[21] For instance, African American children in New York City's West Harlem play at a park built above a massive sewage treatment plant. Improper removal of asbestos from New York's inner-city schools made national headlines in 1993 and caused long delays in the start of the school year.[22]

A similar set of exposures occurred in Dallas, Texas. Soil on the playground at a West Dallas Boys Club, which is in an African American neighborhood that is home to more than 1,200 youth, was so contaminated with lead that outside activities had to be suspended. Health officials discerned that the lead level at the Boys Club was sixty times the level considered potentially dangerous to children. A nearby schoolyard had a similar level of contamination, and a day care facility was forced to close because of the lead problem. Some children in the neighborhood have suffered irreversible brain damage because of the severe and frequent exposure to lead.[23] More than a third of the children in some areas of the community were found to have elevated blood-lead levels.

Nationally, African American children living below the poverty line are exposed to lead levels dangerous enough to cause severe learning disabilities and other neurological disorders at nearly nine times the national rate for more economically advantaged children.[24] Herbert Needleman reports that as many as 55 percent of low-income, African American children have blood-lead levels associated with adverse effects on the nervous system.[25] It has been estimated that under most recent standards, 96 percent of African American children who live in inner cities have unsafe amounts of lead in their blood. Even in families with annual incomes greater than $15,000, 85 percent of African American children in cities are estimated to have unsafe lead levels, compared to 47 percent of white children.[26]

The Location of Hazardous Facilities

Clearly, then, race and income are major factors in the location of hazardous facilities. In most instances, both of these factors come into play. The seminal 1987 study by the Commission for Racial Justice found that three of the five largest commercial hazardous waste facilities in the United States are located in predominantly low-income, Black communities. It also found that three of every five African Americans and Latinos live in communities with uncontrolled toxic waste sites; most have levels of poverty higher than the national average. Similarly, in Detroit, a person of color's chance of living within a mile of a hazardous waste facility is four times greater than a white American's. The proportion of people whose income is below the poverty line is also higher among those residing within a mile of a commercial hazardous waste facility in Detroit.[27]

Almost every study of environmental hazards has concluded that there are racial and income disparities in the location of these facilities. The findings of the subset of studies in table 1, which focused on hazardous waste facilities, support this conclusion (see table 2). All of these studies found racial disparities and all but three discerned income disparities. Only one (Zimmerman, 1993) reported findings that questioned the prevalence of income disparities in the location of environmental hazards. Although pervasive, this disparity in the exposure to environmental hazards is not limited to facility siting decisions.

Table 2. Studies on the Siting of Environmental Hazards

				Disparity	
Year	*Author*	*Type of Hazard*	*Geographic Focus*	*Race*	*Income*
1983	U.S. GAO	Hazardous waste	Southeast	Yes	*
1984	Greenberg and Anderson	Hazardous waste	New Jersey	Yes	Yes
1987	UCC and PDA	Hazardous waste	National	Yes	Yes
1991	Costner and Thornton	Hazardous waste	National	Yes	Yes
1992	Fitton	Hazardous waste	National	Yes	Yes
1992	Ketkar	Hazardous waste	New Jersey	Yes	*
1992	McDermott	Hazardous waste	National	Yes	*
1992	Mohai and Bryant	Hazardous waste	Detroit, Mich.	Yes	Yes
1992	Nieves	Toxic waste/pollution	National	Yes	Yes
1992	Unger et al.	Hazardous waste	Pinewood, S.C.	Yes	Yes
1993	Been	Hazardous waste siting	Southeast	Yes	Yes
1993	Hamilton	Hazardous waste siting	National	Yes	Yes
1993	Zimmerman	Hazardous waste	National	Yes	No

*Income disparities were not addressed in these studies.
Source: Derived from Benjamin A. Goldman, *Not Just Prosperity: Achieving Sustainability with Environmental Justice* (Washington, D.C.: National Wildlife Federation, 1993).

A National Pattern

The poor and people of color are exposed disproportionately to environmental hazards where they live in every region of the country (see table 1). This exposure comes from a variety of sources. In Los Angeles, automobile pollution is worse in low-income African American and Latino neighborhoods. The Latino and African American communities of East Los Angeles, Huntington Park, and Watts in California are also home to metal plating and furniture manufacturing plants that emit toxic chemicals.[28]

The predominantly African American and Latino south side of Chicago boasts the largest concentration of municipal and hazardous waste dumps in the country. One housing project in Chicago is built on an abandoned landfill and surrounded by nine industrial facilities known to emit toxins.[29] Similarly, carbon monoxide from traffic and sulfur dioxide from factories and power plants have been found to reach their highest levels in air in African American and low-income neighborhoods in Washington, D.C.[30]

Native Americans also experience frequent and severe exposure to environmental hazards; many also have incomes at or below the poverty level. Some of the worst toxic pollution problems in the

United States are on their lands. The 1987 Commission for Racial Justice report discovered that approximately half of all Native Americans live in communities with an uncontrolled toxic waste site. Water contamination, uranium mill tailings, chemical lagoons, and illegal dumps are cause for major concern in many of these communities.[31] For instance, Navajo teenagers living in uranium districts suffer from reproductive organ cancers at seventeen times the national rate.

The disproportionate location of environmental hazards in communities where people of color live led the Commission for Racial Justice to conclude:

> The possibility that these patterns resulted by chance is virtually impossible, strongly suggesting that some underlying factor or factors, which are related to race, played a role in the location of commercial waste facilities.[32]

Several attempts have been made to explain why communities where people of color and the poor live are selected more frequently for the siting of environmental hazards. One explanation offered relates to a set of syndrome behaviors.[33]

Syndrome Behaviors and Siting Decisions

Several syndromes prevail that make the siting of environmental hazards in communities where people of color and the poor live politically and economically expedient. Besides the NIMBY (Not in My Backyard) behaviors often discussed in the media, four other syndromes have been discerned:

- NIMEY (Not in My Election Year);
- NIMTOO (Not in My Term of Office);
- PIITBY (Put It in Their Backyard); and
- WIMBY (Why in My Backyard).

The NIMBY syndrome has caused politicians to wither in the face of their constituents. Siting delays associated with this syndrome have been extremely costly for several companies seeking to develop commercial facilities. NIMBYS organize, march, sue, and petition to block developers they think are threatening them. They use the political and legal systems to cause interminable delays. As Richard Andrew found in his study of 179 attempts to site hazardous waste facilities across the United States, of the 25 percent rejected and the 53 percent delayed, NIMBYism was significant in almost every instance.[34] When practiced, NIMBY behavior has resulted in effective campaigns against environmental hazards.[35]

Any mention of environmental hazards usually results in NIMBY behaviors in affluent communities, which in turn, lead to NIMTOO and NIMEY behaviors by elected officials. Pressure for a solution to problems in siting environmental hazards forces these officials to look for a compromise, likely PIITBY.

This PIITBY compromise often results in a decision to place environmental hazards in communities where the poor and people of color live. Circumstances both internal and external to these communities encourage their selection as sites for these facilities. The priority exhibited in site selection is one such circumstance. Principally, sites given the most attention will be those that affect more affluent communities. Such communities have the resources, knowledge, and contacts to sustain the symptoms of the NIMBY syndrome. Accordingly, residents from these communities are more likely to be proactive. They are, generally, the driving force that causes politicians to exhibit both NIMTOO and NIMEY syndrome behaviors.

The NIMBY, NIMTOO, NIMEY, and PIITBY syndromes are seen less frequently in communities where people of color and the poor live. These communities are more prone to exhibit the WIMBY syndrome. That is, they are usually more reactive than proactive in their response to environmental hazards. The WIMBY syndrome emanates from the social, economic, and political realities that often surround poor communities and people of color.

People of color and low-income individuals usually do not have the resources, or contacts, to initiate or sustain the proactive behavior found in more affluent communities. Nor do they have the contacts in government and industry necessary to become involved during the preplanning and planning stages for the siting of environmental hazards. These factors and others have led to a "knowledge and information" gap among people of color about environmental risks.

Perhaps, because of the tradition of having landfills and other waste facilities in their communities, there is also more "social acceptance" for facilities that represent environmental hazards.[36] Thus, the activism or WIMBY syndrome exhibited by low-income individuals and people of color tends to be prevalent after facilities have been constructed or other crucial decisions have been made.

The WIMBY syndrome is, moreover, far more congenial to the NIMTOO and NIMEY political behaviors than the NIMBY syndrome. For elected officials and other politicians, it is safer to investigate "why something was done" than to intervene "while something is being done." The "why" is less likely to affect voter decisions. Most of the politically sensitive decisions will have already been made when symptoms of the WIMBY syndrome become apparent. Decisions about zoning, building permits, and franchise licenses can occur with little or no public outcry.

The greatest concern is often raised by low-income groups and people of color after facilities are operational. Residents then learn that environmental hazards are not just irritants to be tolerated; they pose serious health threats. An examination of health statistics reveals that counties with the worst rank across all of the industrial toxin measures are usually the counties with the worst mortality from all diseases. These are counties with large numbers of people of color; many counties can also be described as economically distressed.[37] In addition to risks that emanate from the siting of environmental hazards and ambient air pollutants in their communities, the poor and people of color are generally more likely to be exposed to environmental hazards through the foods they eat.

Environmental Hazards and Food

The risks that individuals experience from environmental hazards through their food intake is easily illustrated by focusing on accidents at two nuclear facilities. The first facility is the U.S. government's Savannah River nuclear weapons plant in South Carolina. The other is the little-known Peach Bottom nuclear facility in Lancaster, Pennsylvania. Researchers have found an alarming correlation between increases in cancer and infant mortality rates in several African American communities and accidents at these facilities.[38] Radioactive contamination of food products has arisen as the likely medium through which individuals were exposed to environmental hazards that occurred because of these nuclear accidents.

Evidence of possible food contamination from the Savannah plant surfaced after it was learned in 1987 that several serious nuclear accidents had occurred at the facility. These accidents were described as among the worst ever documented, yet they were kept secret for over twenty years. Radiation from the plant appears to have been particularly hazardous for African Americans in the region. For instance, South Carolina's nonwhite male cancer rate rose 35 percent faster than in the rest of the country following the first accident at the nuclear plant. In two states surrounding South Carolina (Georgia and North Carolina), the nonwhite male cancer rate increased 28 percent faster than in the rest of the nation.

Following the accidents, the staple foods consumed by African Americans and the rural poor who live in the South were found to be contaminated. Radiation levels in catfish and bream caught in the Savannah River were more than 100,000 times higher than average for fresh fish in New York City.[39] The radiation concentration in collard greens in the area was fifty times higher than the levels in vegetables in New York City, and it was thirty-three times higher in poultry; grains were forty times more contaminated, and milk contamination was eight times higher near the plant than it was in New York City.[40] Anyone who is familiar with southern cuisine knows that fish, collards, rice, and chicken are the main ingredients for what is affectionately referred to as "Soul Food."

The first tragedy at the Savannah nuclear plant is that the accident happened; still worse, people were not warned about the dangers associated with eating contaminated food. Equally tragic results are thought to have emanated from accidents at Peach Bottom. This nuclear facility is located in a rural milk-producing region of Pennsylvania. Major markets for this milk include Washington, D.C., and Baltimore, Maryland. Milk consumed in these two cities has exhibited some of the highest readings for radiation contamination on the East Coast. Researchers have learned that there was a positive correlation between the distribution of contaminated milk produced in the area surrounding Peach Bottom and high infant mortality rates in Washington and Baltimore, cities with large populations that are poor and mostly people of color. This correlation could be just a coincidence. But the month after the facility was closed—due to negligent behavior—Washington's infant mortality rate dropped to the national norm for the first time since the plant began operating in the mid-1960s.[41]

The possible food contamination from the nuclear facilities mentioned above may not be an everyday occurrence. An important observation, however, is how siting decisions may contribute to contamination in food and its distribution, an area often overlooked by risk assessment studies.

Other evidence suggests that the health of those who are poor and people of color may be routinely at risk from environmental hazards transmitted through the foods they eat. Health risk from contaminated food is generally greater for low-income groups that are mostly people of color than for individuals from other socioeconomic groups. In Detroit, for example, people of color and low-income groups were found to consume the greatest amounts of fish contaminated by municipal and industrial toxins dumped into Michigan's surface water.[42] These findings echo a 1989 report by the Kellogg Foundation, which noted that potentially cancer-causing or nerve-destroying substances, like PCB, now found in many fish are at critical levels in the blood of one-fourth of the children age five and under in some cities.[43]

The disproportionate risk exposure that people of color and low-income groups frequently experience because of food contamination is not, however, limited to the inner cities. Studies of dietary preferences among the Navajo suggest that they also regularly consume food products contaminated with both radiation and lead.[44] The Chippewa take similar risks with their food supply. Mining activities adjacent to their lands threaten toxic contamination of the fish, deer, and wild rice that make up a major portion of the food supply for the Chippewa.

Reasons for Food Contamination Disparities

Why are the poor and people of color more at risk of consuming toxins in the foods that they eat? Explanations have been offered for this disparity that include a malicious conspiracy, the profit motive, and cultural insensitivity. The first two explanations have been widely discussed by scholars and others seeking to address the misery often inflicted by racism and unregulated capitalism. The malicious conspiracy theory purports that there has been a deliberate attempt to cover up the fact that foods frequently consumed by the poor and people of color contain higher levels of carcinogens than those consumed by other groups. According to Jay Gould and Benjamin Goldman, in their book *Deadly Deceit,* there has been a governmental cover-up that appears to have included outright falsification of data.[45]

In addition to the cover-up asserted by Gould and Goldman, Bullard argues that the malicious conspiracy also includes environmental blackmail. That is, people of color, because of economic constraints, are forced to accept circumstances and conditions that may be hazardous to them, their families, and their communities.[46] This includes consuming food that may contain toxins.

The profit motive theory emanantes from, but is not limited to, the Marxist analysis of environmental justice issues. Proponents of this explanation argue that environmental concerns represent a threat to the earning power of capitalists.[47] Hence, the environmental injustice inflicted on people of color and the poor has become a mechanism for recouping profits that have been lost because of environmental regulations. As Ivette Perfecto contends, environmental injustice inflicted on people of color and the poor is another form of expropriation permitted under capitalism:

> Expropriation is allowed to occur either (1) because it is hidden from view from those expropriated (the marvel of the capitalist system of class exploitation), or (2) because those expropriated have no right or political power to resist the expropriation. Race and the environment thus can be formulated as two sides of the same coin.[48]

The profit motive theory suggests that the falsification of data and economic blackmail that allow people of color and the poor to be more at risk of having toxins in the foods they eat are merely forms of expropriation used by the capital-owning classes. It also represents an important attempt at intersecting race and class variables.

Discussed less frequently is the cultural insensitivity theory for environmental injustice. This explanation suggests that the disproportionate risk experienced is, at least in part, the result of policies designed to protect the public. That is, these policies are usually based more on the dietary preferences and eating habits of select groups of European Americans. For instance, the Detroit fish case, mentioned above, involves policymaking that uses an average consumption rate for the state of Michigan in its standards-setting process. This process does not consider the variations in the levels of consumption by subgroups of the Michigan population. Thus, because of the lack of cultural sensitivity, the disproportionate risk experienced by people of color, the rural poor, and other low-income groups is said to often be the result of benign neglect.

Whether the result of overt or covert racism, putting economic profits over the health of people, or benign neglect, this disproprotionate risk can and does lead to disastrous results. An injustice exists even if it is merely a coincidence that:

- the food, air, and water that people of color and those who are poor consume are more contaminated;
- nonwhite workers are 50 percent more likely to be exposed to hazards in the workplace; and
- hazardous waste facilities are located disproportionately in communities where people of color and the poor live.

A society that allows such a pattern of coincidences to persist has failed to equally protect its citizens. This failure, itself, constitutes an environmental injustice.

Environmental Justice as a Policy Concern

Environmental justice received official recognition as a federal policy concern in the United States in 1993 when the Clinton administration established by executive order the President's Council on Sustainable Development. This concern was further recognized in 1994 with the signing of Executive Order 12898, which created the National Environmental Justice Advisory Council to the EPA. The action taken by the president calls for the development of environmental justice strategies throughout a number of organizations. For instance, the executive order mandates: (1) the coordination of government agencies in addressing environmental justice problems, and (2) the support of grassroots community participation in human health research, including data collection and analysis where practical and appropriate.

Table 3. Principles of Environmental Justice

1. Environmental justice affirms the sacredness of Mother Earth, ecological unity and the interdependence of all species, and the right to be free from ecological destruction.
2. Environmental justice demands that public policy be based on mutual respect and justice for all peoples, free from any form of discrimination or bias.
3. Environmental justice mandates the right to ethical, balanced and responsible uses of land and renewable resources in the interest of a sustainable planet for humans and other living things.
4. Environmental justice calls for universal protection from nuclear testing, extraction, production and disposal of toxic/hazardous wastes and poisons and nuclear testing that threaten the fundamental right to clean air, land, water, and food.
5. Environmental justice affirms the fundamental right to political, economic, cultural, and environmental self-determination.
6. Environmental justice demands the cessation of the production of all toxins, hazardous wastes, and radioactive materials, and that all past and current producers be held strictly accountable to the people for detoxification and containment at the point of production.
7. Environmental justice demands the right to participate as partners at every level of decision making including needs assessment, planning, implementation, enforcement, and evaluation.
8. Environmental justice affirms the right of all workers to a safe and healthy work environment, without being forced to choose between an unsafe livelihood and unemployment. It also affirms the right of those who work at home to be free from environmental hazards.
9. Environmental justice protects the rights of victims of environmental injustice to full compensation and reparations for damage as well as quality health care.
10. Environmental justice considers governmental acts of environmental injustice a violation of international law, the Universal Declaration on Human Rights, and the United Nations Convention on Genocide.
11. Environmental justice must recognize a special legal and natural relationship of Native Peoples to the U.S. government through treaties, agreements, compacts, and covenants affirming sovereignty and self-determination.
12. Environmental justice affirms the need for urban and rural ecological policies to clean up and rebuild our cities and rural areas in balance with nature, honoring the cultural integrity of all our communities, and providing fair access for all to the full range of resources.
13. Environmental justice calls for the strict enforcement of principles of informed consent, and a halt to the testing of experimental reproductive and medical procedures and vaccinations on people of color.
14. Environmental justice opposes the destructive operations of multinational corporations.
15. Environmental justice opposes military occupation, repression and exploitation of lands, peoples and cultures, and other life forms.

Table 3. *Continued*

16. Environmental justice calls for the education of present and future generations which emphasizes social and environmental issues, based on our experience and an appreciation of our diverse cultural perspectives.
17. Environmental justice requires that we, as individuals, make personal and consumer choices to consume as little of Mother Earth's resources and to produce as little waste as possible; and make the conscious decision to challenge and reprioritize our lifestyles to insure the health of the natural world for present and future generations.

Source: The First People of Color Environmental Leadership Summit, Washington, D.C., 27 October 1991.

Although significant, the president's administrative measures constitute only one portion of the effort that has made environmental justice a policy concern. Legislative initiatives at the state and federal levels have also helped to focus attention on the strong evidence supporting charges that people of color and the poor suffer disproportionately from environmental hazards. As a result, legislation has been initiated in ten states and at least five bills have been introduced in the U.S. Congress that address environmental justice concerns.[49] These initiatives are partially a response to the mounting body of evidence on the disproportionate impact of environmental risks, but they are also the result of grassroots organizing efforts by individuals affected by these hazards.

Hundreds of individuals involved in these struggles met in October 1991 at the First National People of Color Environmental Leadership Summit, where they outlined seventeen "Principles of Environmental Justice" (see table 3). These principles declare clean air, land, water, and food to be a fundamental right. They affirm the right of all workers to a safe and healthy environment. The right to participate as equal partners at every level of decision making is also demanded.

Although participants at the summit included some of the leading researchers contributing to the body of literature on environmental justice, the principles were formulated by members of grassroots and indigenous organizations. The principles call for the strict enforcement of informed consent, and a halting to the testing of experimental reproductive and medical procedures and vaccinations on people of color. Though speculative at the time, this admonition now seems appropriate with recent revelations that approximately 9,000 Americans, including children and newborns, were used in 154 human radiation tests sponsored by the U.S. Department of Energy. Many of those tested were people of color and the poor.[50]

A diverse body of research and researchers suggests that low-income individuals, particularly those of color, have been asked to bear a disproportionate share of the burdens associated with environmental hazards. The result has been the elevation of environmental justice to the policy level. Prominent individuals and organizations have become champions for environmental justice. Nevertheless, people of color and low-income groups continue to experience more frequent and severe exposure to risks from environmental hazards than do other groups in U.S. society.

Notes

1. Benjamin A. Goldman, *Not Just Prosperity: Achieving Sustainability with Environmental Justice* (Vienna, Va.: National Wildlife Federation, 1993).
2. Benjamin A. Goldman, *The Truth about Where You Live: An Atlas for Action on Toxins and Mortality* (New York: Time Books/Random House, 1991).
3. See Philip J. Landrigan, "Prevention of Toxic Environmental Illness in the Twenty-first Century," *Environmental Health Perspectives* 86 (1990): 197–99; and Richard Doll and Richard Peto, *The Causes of Cancer: Quantitative Estimates of Avoidable Risks of Cancer in the United States Today* (New York: Oxford University Press, 1981).
4. As reported Goldman, *Truth about Where You Live.*
5. Sue Pollack and JoAnn Grozuczak, *Reagan, Toxics, and Minorities: A Policy Report* (Washington, D.C.: Urban Environment Conference, 1984).
6. Anthony J. McMichael, "Mortality among Rubber Workers: Relationship to Specific Jobs," *Journal of Occupational Medicine* 18 (1976): 178–84.
7. Larry Williams, "Long-Term Mortality of Steelworkers; I: Methodology," *Journal of Occupational Medicine* 11 (1969): 301.
8. Beverly H. Wright, "The Effects of Occupational Injury, Illness, and Disease on the Health Status of Black Americans: A Review," in *Race and the Incidence of Environmental Hazards: A Time for Discourse,* ed. Bunyan Bryant and Paul Mohai (Boulder, Colo.: Westview Press, 1992).

9. William Blot, J. Malcolm Harrington, Ann Teledo, Robert Hoover, Clark Heath Jr., and Joseph F. Fraumeni Jr., "Lung Cancer after Employment in Shipyards during World War II," *New England Journal of Medicine* 299 (12): 620–24.

10. Robert F. Wasserstrom and Rose Wiles, *Field Duty, U.S. Farmworkers, and Pesticide Safety,* study 3 (Washington, D.C.: World Resources Institute, Center for Policy Research, 1985).

11. Ivette Perfecto, "Pesticide Exposure of Farm Workers and the International Connection," in Bunyan and Mohai, *Race and the Incidence of Environmental Hazards.*

12. See Marion Mosses, "Pesticide-Related Health Problems and Farmworkers," *American Association of Occupational Health Nurses Journal* 37 (1989): 115–30; Valerie A. Wilk, *The Occupational Health of Migrant and Seasonal Farmworkers in the United States* (Washington, D.C.: Farmworkers Justice Fund, 1986); Pollack and Grozuczak, *Reagan, Toxics, and Minorities;* and E. Kahn, "Pesticide Related Illnesses in California Farmworkers," *Journal of Occupational Medicine* 18 (1976): 693–96.

13. As reported in Perfecto, "Pesticide Exposure."

14. Nora Lapin and Karen Hoffman, *Occupational Disease among Workers: An Annotated Bibliography* (Hyattsville, Md.: National Institute for Occupational Safety and Health, 1981).

15. Goldman, *Truth about Where You Live.*

16. Prominent among these Bryant and Mohai, *Race and the Incidence of Environmental Hazards;* Goldman, *Truth about Where You Live;* Commission for Racial Justice, *Toxic Wastes and Race: A National Report on the Racial and Socioeconomic Characteristics of Communities with Hazardous Wastes Sites* (New York: United Church of Christ, 1987); Robert D. Bullard, *Dumping in Dixie: Race, Class, and Environmental Quality* (Boulder, Colo.: Westview Press, 1990); and U.S. General Accounting Office, *Siting of Hazardous Waste Landfills and Their Correlation with Racial and Economic Status of Surrounding Communities* (Washington, D.C.: General Accounting Office, 1993).

17. Goldman, *Truth about Where You Live.*

18. Bullard, *Dumping in Dixie.*

19. Martha Allen, "Asbestos in Chicago Housing Authority Apartments Poses Possible Health Hazards," *Chicago Reporter* 15 (1986): 1–4; and Herbert Needleman, "Childhood Lead Poisoning: Mandate and Eradicable," *American Journal of Public Health* (1991): 685–87.

20. Needleman, "Childhood Lead Poisoning"; and Beverly H. Wright, "Environmental Equity Justice Centers: A Response to Inequity," in *Environmental Justice: Issues, Policies, and Solutions,* ed. Bunyan Bryant (Washington, D.C.: Island Press, 1995).

21. Bullard, *Dumping in Dixie;* Goldman, *Not Just Prosperity;* and Matthew L. Ward, "Fear of Asbestos in the Air," *New York Times,* 26 September 1993.

22. Steven L. Myers, "Officials Admit Asbestos Deadline Was Unrealistic," *New York Times,* 3 September 1993.

23. Wasserstrom and Wiles, *Field Duty.*

24. John F. Rosen, "Metabolic Abnormalities in Lead Toxic Children: Public Health Implications," *Bulletin of the New York Academy of Medicine* 10 (1989): 1063–83.

25. Needleman, "Childhood Lead Poisoning."

26. Wright, "Environmental Equity"; and Karen L. Florini, *Legacy of Lead: America's Continuing Epidemic of Childhood Lead Poisoning* (Washington, D.C.: Environmental Defense Fund, 1990).

27. Bryant and Mohai, *Incidence of Environmental Hazards.*

28. Eric Mann, *LA's Lethal Air: New Strategies for Policy, Organizing, and Action* (Los Angeles, Calif.: Labor/Community Strategy Center, 1991).

29. Goldman, *Truth about Where You Live.*

30. Julian McCaull, "Discriminatory Air Pollution: If Poor, Don't Breath," *Environment* 2 (1976): 26–31.

31. Chris Shuey, "Uranium Mill Tailings: Toxic Waste in the West," *Engage/Social Action* (October 1984): 40–45.

32. Commission for Racial Justice, *Toxic Wastes and Race,* 23.

33. Harvey L. White, "Hazardous Waste Incineration and Minority Communities," in Bryant and Mohai, *Race and the Incidence of Environmental Hazards.*

34. As reported in Walter A. Rosenbaum, *Environmental Politics and Policy* (Washington, D.C.: CQ Press, 1991).

35. Patrick G. Marshall, "Not in My Backyard," *CQ Editorial Reports* (Washington, D.C.: Congressional Quarterly, June 1989).

36. Michael R. Edelstein, *Contaminated Communities: The Social and Psychological Impact of Residential Toxic Exposure* (Boulder, Colo.: Westview Press, 1988).

37. Goldman, *Truth about Where You Live.*

38. Jay M. Gould and Benjamin A. Goldman, *Deadly Deceit: Low-Level Radiation, High-Level Coverup* (New York: Four Walls Eight Windows, 1991).

39. Ibid.

40. Ibid.

41. Ibid.

42. P. C. West, "Minorities and Toxic Fish Consumption: Implica-

tions for Point Discharge Policy in Michigan," in Bryant, *Environmental Justice;* and P. C. West, J. M. Fly, F. Larkin, and R. W. Marans, "Minority Anglers and Toxic Fish Consumption: Evidence from a State-wide Survey of Michigan," in Bryant and Mohai, *Incidence of Environmental Hazards.*

43. Joseph D. Beasley and Jerry J. Swift, *The Kellogg Report: The Impact of Nutrition, Environment, and Lifestyles on the Health of Americans* (Annandale-Hudson, N.Y.: Institute of Health Policy Practice, 1989).

44. Bryant and Mohai, *Incidence of Environmental Hazards.*

45. Gould and Goldman, *Deadly Deceit,* 73.

46. Bullard, *Dumping in Dixie,* 83.

47. See, for example, Perfecto, "Pesticide Exposure"; and Ivette Perfecto, "Sustainable Agriculture Embedded in a Global Sustainable Future: Agriculture in the United States and Cuba," in Bryant, *Environmental Justice.*

48. Perfecto, "Pesticide Exposure," 178–79.

49. Goldman, *Not Just Prosperity.*

50. Karen Mcpherson, "Clinton Regrets Radiation Tests," *Albuquerque (N.M.) Tribune,* "Human Radiation Tests," 4 October 1995, A-6.

Jeanne Nienaber Clarke and Andrea K. Gerlak

ENVIRONMENTAL RACISM IN SOUTHERN ARIZONA? The Reality beneath the Rhetoric

Southern Arizona is a land of much cultural diversity. Native Americans, Latinos, African Americans, Asian Americans, and Anglo-Americans live in close proximity with one another. These five ethnic groups make up the population of metropolitan Tucson, which, with its 405,000 residents, is the demographic center of Pima County. Anglos are the most numerous, comprising about 65 percent of the population. Latinos account for 29 percent, African Americans 4 percent, Native Americans constitute 2 percent, and Asians account for less than 1 percent.[1] All share the same local governing structure: a five-member Pima County Board of Supervisors, and a mayor and five-member council for the city of Tucson. These two elected bodies are independent of one another, but needless to say, there exists considerable overlapping of functions between city and county governments. In addition, the region's principal tribal government is the Tohono O'odham (formerly known as the Papago).

Metropolitan Tucson is situated in the Sonoran Desert and is surrounded by mountain ranges. The extensive federal landholdings that circle the city effectively limit growth to the valley floor. But because it is a large basin, stretching nearly fifty miles in both directions, there has always been ample room for growth.

And growth there has been. Like any number of Sun Belt cities, Tucson's population—especially its Anglo one—grew along with the spread of air-conditioning and other modern amenities. From a town of about 38,000 at the end of World War II, Tucson grew to become, in 1992, the nation's thirty-fifth largest city.[2] Tucsonans have accommodated this substantial population increase about as well as residents in other Sun Belt cities: they have experienced the usual growing pains and have dealt with them in the customary manner. Heated debates over growth have kept local officials busy for the last twenty years as Tucson's population doubled. The debates, however, have had a negligible impact on stemming the population tide.

Tucson is, then, a typical postwar city, built largely on the technologies and industries that developed during and after World War II. With its Davis Monthan Air Force Base and Hughes Aircraft Company (formerly known as Air Force Plant 44), the city has a pronounced military presence. As the scholar Barry Commoner has contended,[3] and as discussed here, Tucson's most serious pollution problem is the result of decades of industrial activity that employed synthetic chemical compounds. More particularly, it was the "high-tech" research and development sponsored by the Air Force and carried out by Hughes Aircraft that produced a Superfund site in south Tucson in the 1980s.

At first glance, Tucsonans all share the same physical environment. The groups and cultures that make up the population are bunched together in the valley floor. They share the same land, the same climate, the same water, the same air. What affects one group affects the other groups. Although it has long been recognized that the population of Pima County is culturally diverse, it was assumed that its residents, at the minimum, shared a physical space. Therefore, they were capable of pulling together to solve some common problems, such as basinwide air- and water-borne pollution.

But do they, in reality? Are there not identifiable pockets of pollution existing *within* a metropolitan area? Are some groups environmentally worse off than others? This chapter utilizes Robert Bullard's definition of environmental racism to analyze the situation in metropolitan Tucson:

> Ecological inequities in the United States result from a number of factors, including the distribution of wealth, housing and real estate practices, and land use planning. . . . Taken together, these factors give rise to what can be called "environmental racism": practices that place African Americans, Latinos, and Native Americans at greater health and environmental risk than the rest of society.[4]

We examined this proposition using data from several sources: in-depth interviews with the five elected members of the Pima County Board of Supervisors and with several other opinion leaders in Tucson; documents pertaining to pollution impacts collected by Supervisor Raul Grijalva since his election to the board in 1988; and four maps produced by Pima County's Departments of Environmental Quality and Public Works showing the location of environmental impacts, the board of supervisors' districts, the location of minority groups, and the location of low-income residents.[5]

The conclusions concerning the validity of the environmental racism concept follow from an examination of these data. This chapter attempts to explain why, for example, one elected official claimed that the concept is based on "a thin fabric of reality," while another elected official is convinced that environmental racism is a feature of local politics. We also address a key issue in the debate over environmental racism—whether local land use decisions are economically or racially driven, or both.

Who Thinks What and Why

To the question, Does environmental racism exist in Pima County? we received a wide range of answers from elected officials and the other respondents. With respect to the governing board, two of its five members were convinced that Latinos and other people of color were affected disproportionately by pollution. Two others believed that the concept had virtually no validity. The fifth member thought that there might be some truth to it, especially with respect to actions taken in the past. He also felt that the term was divisive and confrontational, and therefore, that it should be discarded.

Clearly, there is little agreement among the five elected officials of Pima County on whether this is a real or an imagined issue. Moreover, they saw their views as representative of the residents in their districts. Hence, it is reasonable to assume that there is also no consensus on this issue among the population at large; that residents in the affluent Catalina foothills maintain one perspective, while residents of the lower-income Sunnyside and westside neighborhoods believe another.

Ethnicity and party affiliation correlate with views of board members on environmental racism in Pima County. The two Latino members, both native Tucsonans and both Democrats, argue that they know that this condition exists because they have lived with it.

Supervisor Grijalva, in fact, became the first local elected official to publicly discuss political decisions in these terms. He did so shortly after assuming office in 1989. The other three board members are of Anglo background and are affiliated with the Republican Party. None of them are native Tucsonans, although two of the three have lived in Tucson for decades. The youngest board member, who moved to Tucson ten years ago and who is originally from southern California, held the least firm opinions on the issue. He tried, he said, to see both sides. All three men, however, held quite distinct views from those of the two Latino supervisors.

With respect to the environmental racism issue, the two most frequently discussed problems in the Tucson area are water pollution and landfill siting. In these discussions, respondents differ radically in their perceptions of cause and effect, and of what is factually correct and what is not. A synopsis of each issue follows.

Personal Knowledge versus Statistical Analysis

Tucsonans call it the "TCE plume" and it extends for six miles in a south-north direction. It is about one and a half miles wide. Although it is clearly the most serious, and most widely recognized, pollution problem in Pima County, it is not directly observable. The "plume" refers to the contamination of a part of the underground aquifer by the industrial solvent trichloroethylene (TCE). Beginning in 1951, large amounts of TCE and other chemicals were dumped—untreated—in desert arroyos on the southside of the city. This method of disposal by Hughes Aircraft and other industries in the area continued until 1977, when Hughes installed a treatment system.[6]

Because of lax or nonexistent environmental regulations, and also because the contamination was invisible (it was occurring slowly underground), polluting activity went unremarked for nearly thirty years. It was only in 1981, shortly after the passage of more stringent federal antipollution laws, that Tucson water officials started closing wells on the city's southside. Officials did not inform the nearby residents why they were closing them.

Regardless of the fact that local officials had no comment on the well closings, by this time residents of the affected area, who are primarily Latino, knew that something was wrong. "We spoke about the high rates of illness to ourselves," Grijalva explained recently.[7] Those discussions went on for several years and were limited largely to the Hispanic community. Then, in 1985, Jane Kay, an investigative reporter for a local newspaper, did a series of articles on TCE pollution. These articles finally brought the problem to the attention of the entire city. Kay's investigation provided further documentation for what residents knew from firsthand experience: People were getting sick and dying "from having drunk, bathed in, and inhaled—from the use of their evaporative coolers—TCE for twenty-five years," neighborhood activist Manuel Herrera said.[8]

At this time, some Sunnyside residents began thinking that "environmental racism" was a factor in, *if not the polluting activity itself,* the official response to TCE. It is critical to note that charges of racism developed primarily because Pima County's health director, an Anglo female physician, relied on the available epidemiological data to conclude that there was no connection between the incidence of serious illnesses among southside residents and their exposure to TCE. During a meeting in 1985 held at the neighborhood library, the issue was joined when the health director placed the blame for illness and death on the victims' own behavior. She told the largely Latino audience that their diet was bad, they smoked, drank excessively, and didn't exercise enough. "I really feared for her life that night," Tucson attorney Richard Gonzales (who eventually litigated the case) said.[9]

Since that stormy night in 1985, the spokespersons for Tucson's Latino community have been convinced that the water that they and their neighbors have had to drink was more polluted than what the Anglo majority ingested over the same period of time. Anglo opinion leaders tend not to believe this. The debate turns on how to establish cause and effect. The former county health director, and others, relied on a study done in the mid-1980s by the epidemiological branch of the Arizona Department of Health Services (ADHS). That study did not find a causal connection between TCE-contaminated water and higher rates of illness. A 1991 government report cited the ADHS study, observing: "With minor exceptions, no remarkable hot spots were found. . . . Any efforts to convince [EPA] scientists or others that further studies are in order must contend with this ADHS clean bill of health for the neighborhood."[10]

This conclusion was also shared by Supervisor Grijalva's predecessor on the board, an individual whose politics, age, gender, and socioeconomic status are (or were) practically identical with those of Grijalva's. On most issues, he could be counted on to further the interests of his lower-income constituents and, as a member of the board through the 1980s, he distinguished himself as an outspoken reformer. However, he is not Latino and did not live in the affected area.

The former supervisor, who was a board member when the TCE controversy erupted in 1985, did not agree with his Latino, south side constituents about what was causing their health problems. In a 1994 interview, he said that while he felt that environmental racism was "a very useful concept," he also believed that the county health director, who became so unpopular that she had to resign, was made into a "scapegoat" in the TCE conflict.[11] Like the health director, he thought that smoking and other unhealthy "lifestyle" habits were causing premature deaths and the appearance of unusual diseases among the southside residents. When asked how he explained the family history of, for example, Herrera—whose wife, all but one of his children, and a number of grandchildren have had severe physical problems—he replied: "I don't know. Some unusual ailments run in certain families."

The Latino community, by contrast, has relied on its personal knowledge[12] to establish causality. They saw themselves, their families, and their neighbors becoming ill at relatively early ages. Many even died from cancers and rare immune system disorders. But what really incensed the southside residents was that "once the problem did become known, not enough was done. The city is *still* pumping treated water with traces of TCE back to these same homes," one of the Latino supervisors said. "That is unconscionable," Eckstrom remarked, noting that this is precisely where racism enters into the issue.[13] City officials wouldn't dare do the same thing in more affluent, white neighborhoods, he argued.

In 1987, frustrated by what appeared to be a casual and even callous response on the part of local and state officials, 1,600 south side residents brought suit against Hughes Aircraft (and the Air Force). Approximately 70 percent of the plaintiffs were Latinos,[14] despite the fact that they only account for about 30 percent of the population of metropolitan Tucson. The 1991 out-of-court settlement for $85 million "was the largest settlement over water pollution in U.S. history," one of the plaintiffs' attorneys, himself a Latino, stated. "It was a great victory for the community," Gonzalez said. "It empowered us. The litigation had a tremendous impact on the way government looks at us."[15]

The records in this case are sealed because a related suit, this one against the City of Tucson's insurers, is still being litigated. Therefore, the precise reasons behind the settlement are not known. Given the record-breaking size of the settlement, however, it seems likely that the plaintiffs' complaints were found to be valid.

The settlement represented not only a significant political victory for the Latino community but a victory for one form of empirical evidence over another. The *case histories* provided by residents in the area of disease and death overwhelmed *statistics* showing weak or nonexistent correlations between illnesses and the location of TCE-laced water in Tucson.[16] That there exist identifiable differences in the types of information that various ethnic, or cultural, groups rely on to further their political interests is thus seen in this environmental issue. Most people trust what political scientist Aaron Wildavsky calls their "sense of local rationality"[17] to decide what is safe and what is risky, and which actions to support and to oppose. Rationality itself is, to a significant degree, culturally determined.

Fighting Garbage Dumps versus Golf Courses

Unlike water pollution, garbage dumps and landfills are a far more visible environmental problem. It is safe to say that no one wants one in her or his backyard. Nevertheless, the Tucson metropolitan area, like most growth areas in the United States, must have new landfills. The issue is where to put them: or, how far is far enough away, and how far is too far away? The politics of landfill siting could be called the Goldilocks syndrome, because finding the "just right" location takes significant and sustained effort.

For the past several years, this issue has been on the agenda of the Pima County Board of Supervisors. The selection process went on for years before planners narrowed it down to three sites: Green Valley, Rita Ranch, and the airport. While all three are located south of the city, Green Valley is thirty miles away and in the vicinity of a middle-class, mostly white, fast-growing retirement community.

Transportation costs were the major factor, Supervisor Paul Marsh explained, in eliminating the Green Valley site. Then, "We heard a big roar from neighbors in the area of the Rita Ranch Site," he said. Moreover, he added, there are high tension wires and natural gas pipelines in the Rita Ranch vicinity that could be affected by the landfill.[18]

That left the airport site, which was being strenuously pushed by one of the Anglo supervisors. A bitter dispute ensued, pitting the Latino and Anglo board members against each other. The Sunnyside Neighborhood Coalition, which had been formed over the TCE issue, also mobilized to defeat the selection of the airport site. It was too close to their neighborhood and, therefore, represented another case of dumping on the poor, Latino leaders claimed.

"It is a perfect example of environmental racism," said Eckstrom, whose district includes the site.[19] The other Latino board member concurred, explaining that "from a scientific viewpoint, the Rita Ranch Site was the best one."[20] But Rita Ranch, as well as Green Valley, was located in the district of an Anglo supervisor, and *his* constituents fought successfully to keep it away from their neighborhoods.

With the proposed airport landfill site about two miles away from some residential neighborhoods, the Latino board members argued that it was too close to *their* homes. Moreover, according to activist Herrera, the site is located over a watershed that runs southeast to northwest. (One end of the TCE plume is located a few miles northwest of the site.) Herrera said that a third objection to the site involved plans being made by the Sunnyside School District to build an elementary school in the vicinity. Thus, for all of these reasons, the Latino community vigorously fought the airport site.

"This is a non-issue," Mike Boyd remarked. "The airport site is so far away from the nearest population" that it makes no sense to allege environmental racism in the selection of the landfill site.[21] Indeed, all three Anglo supervisors agreed on that point: one supervisor said that there were no homes in the area and that the closest homes were 2⅜ miles away; and another said that the airport site was 2½ miles away from the nearest residences. He went on to point out that people in fashionable Scottsdale, Arizona, live much closer than that to a landfill, as they do in the small town of Marana. "Economics, pure and simple, drove the decision," Ed Moore said.[22] It had nothing to do with environmental racism.

Clearly, opinions about the two-mile distance lie in the eye of the beholder. For those living in the vicinity, it is too close; for others, it is not. Moreover, Tucsonans' interest in the landfill siting issue varies in inverse proportion to their distance from it. "Nine out of ten of my constituents wouldn't have cared where they put it—so long as it wasn't in their [my] district," a northside Anglo supervisor claimed.[23]

Indeed, the Latino supervisor in whose district the airport site is located was especially bitter about the white majority's lack of interest. "Where were the middle-class environmentalists when it came to fighting the Airport Site?" he asked. The "saguaro-savers" get up in arms over plans to build another golf course in the foothills, but they don't come, he said, to hearings and meetings about garbage dumps on the southside.

"This is the racist part of the whole thing," Eckstrom claimed.[24] Because the established environmental organizations weren't there on the landfill issue, just as they weren't involved in the southside's water pollution problem, he saw little likelihood of a broad-based coalition emerging among environmental activists in southern Arizona. "We had to plead with the Sierra Club and the Audubon Society just to get letters supporting our position on the TCE issue," the other Latino supervisor said. "And they are supposed to be 'advocacy' groups," Grijalva noted.[25]

In action taken at the end of 1994, the Board of Supervisors killed all three proposed landfill sites. The siting process began de novo in 1995 and a site still had not been selected as of the fall of 1996.[26] Allegations of environmental racism made by Latino supervisors and others contributed to the decision to wipe the slate clean and start over. "It definitely represented a success for the southside residents," Grijalva said. Without the immediacy of the TCE issue—and without the new, racially charged rhetoric—it is likely that the airport site would have been selected for a new regional landfill.

While this is perceived as a victory by and for minority residents in Pima County, others view it differently. The longer it takes to find an acceptable location for what all agree is a necessity, the more expensive it will be for the community as a whole. Already the days of free garbage pickup within city limits—which city officials have used to attract businesses and jobs to Tucson—appear to be numbered. They will be a casualty in the NIMBY (Not in My Backyard) wars that increasingly define the landscape of local politics, and which more and more resemble a two-edged sword.

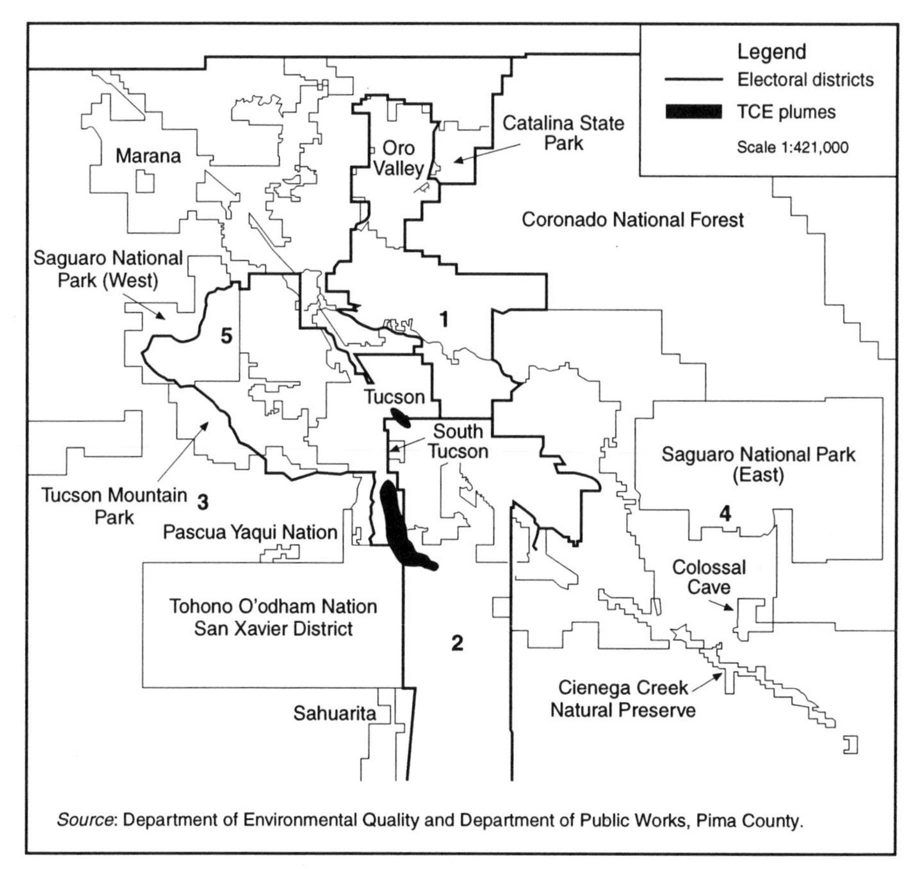

Map 1. Pima County Board of Supervisors Districts, Metropolitan Tucson Area, October 1997

Mapping Political Boundaries, Environmental Impacts, Race, and Poverty

Each member of the Pima County Board of Supervisors represents a geographic district. Map 1 shows the boundaries of the five electoral districts. Districts one, three, and four currently are represented by the three Anglo, Republican members, and districts two and five are represented by the two Latino, Democratic members.

Map 2, produced in 1997 by the Pima County Department of Environmental Quality, pinpoints the major environmental impacts in the Tucson metropolitan area. These include the large TCE

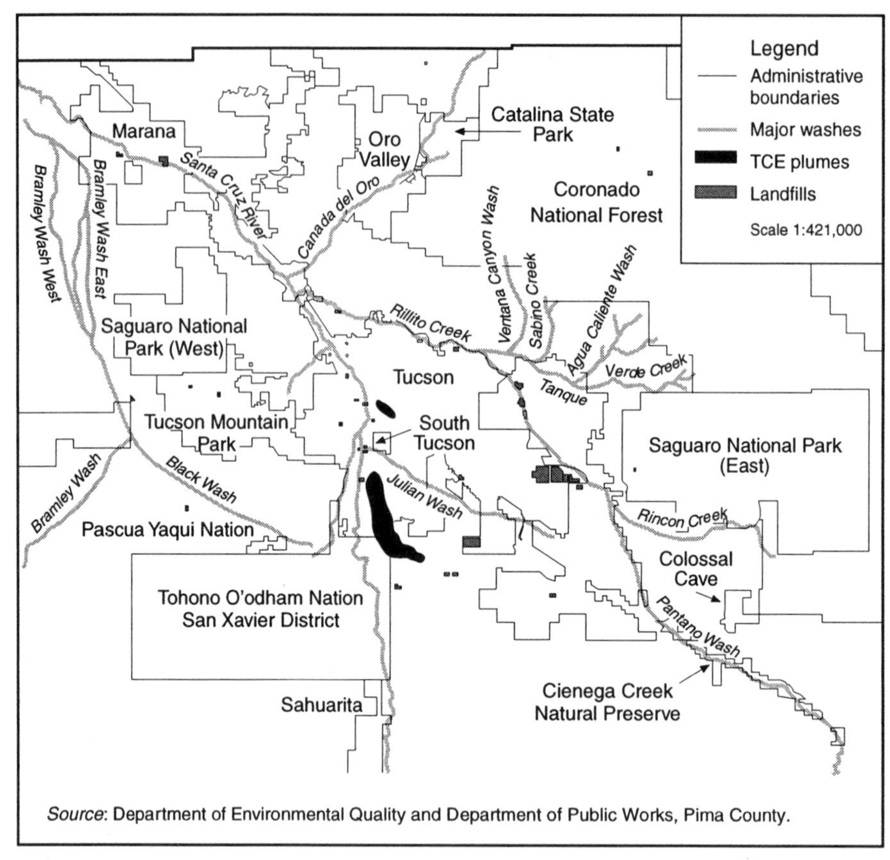

Map 2. Major Environmental Impacts, Metropolitan Tuscon Area, October 1997

plume, another smaller plume of contaminated water to the north of the TCE one, the locations of underground—and leaking underground—storage tanks, and Resource Conservation and Recovery Act (RCRA) facilities. The proposed airport landfill site is not shown on the map, but it is located just to the southeast of the southern end of the TCE plume.

An overlay of map 2 on map 1 indicates that the most heavily impacted districts—two and five—are those of the two Latino supervisors. These districts include the downtown area of Tucson, the small city of South Tucson, part of the Santa Cruz Wash, Interstates 10 and 19 (which bisect the city), the airport, and the near south and west sides. The maps show that most of the industrial corridor of the city is located in districts two and five.

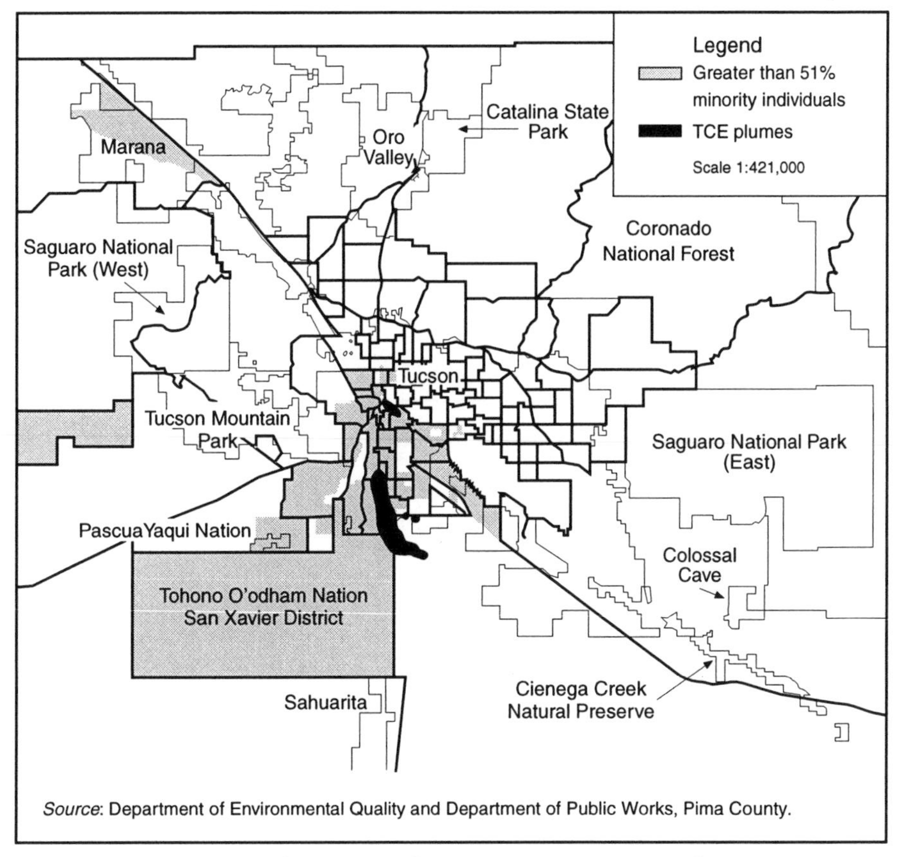

Map 3. Location of Minorities by Census Tract, Metropolitan Tucson Area, December 1994

Map 3, produced for this chapter by the Pima County Department of Environmental Quality and the Department of Public Works, shows the location of minority residents by census tract in metropolitan Tucson. The racial pattern is clear. With the exception of part of the town of Marana to the northwest, the non-Anglo population is concentrated on the south and west sides of the city. The self-governing Tohono O'odham Reservation (the San Xavier district on the map) lies in district three. Most of Tucson's 30 percent Latino population and its small African American community reside in districts two and five. An overlay of map 2 on map 3 shows that the severest environmental impacts, such as the TCE plume, are located in neighborhoods of color.

Map 4, also produced for this chapter, identifies by census tract

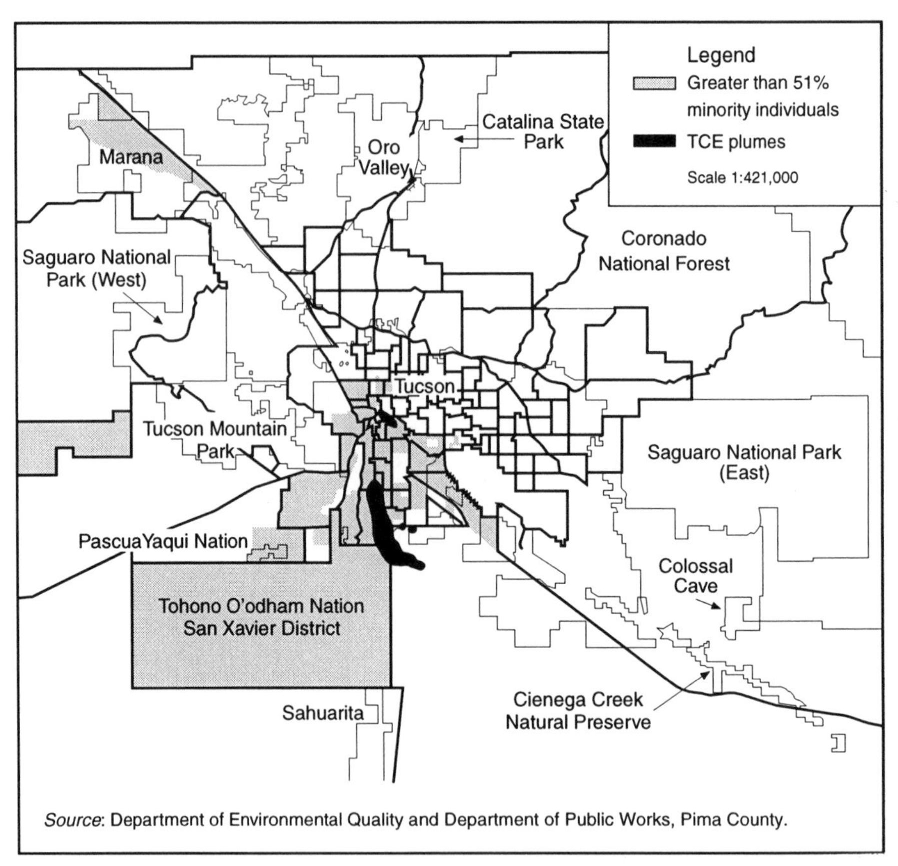

Map 4. Low-Income Households by Census Tract, Metropolitan Tucson Area, October 1997

where the poorest residents of Tucson live: on the Tohono O'odham Reservation and in a narrow corridor stretching for several miles to the north and south from the city's downtown. The physical concentration of low-income households is stark. Once again, most of Tucson's poor are located in districts two and five, and within the city's industrial sector.

An overlay of these four maps further explains why the five supervisors hold the opinions that they do on the issue of environmental racism. The districts of the two Latino supervisors are ethnically, economically, *and* environmentally distinct from the other three districts. Moreover, the political geography of metropolitan Tucson closely resembles what another researcher, using geographic information systems software, reported to be the case for metropolitan Los Angeles.[27] Complaints about environmental inequities have been raised by minority leaders in both cities precisely because of the "overlay effect" of ethnicity, poverty, and the existence of pockets of pollution.

As does Bullard's pathbreaking work, this research into the geopolitics of pollution finds that economics and race are inextricably intertwined. Those scholars who attempt to isolate economics from racism as causal factors in explaining environmental inequity, therefore, are missing the point. In fact, such efforts to tease out, for analytical purposes, the effects of each of these discrete variables on pollution impacts can itself be seen as a form of racism. Certainly, from the perspective of people of color having to deal with a dirtier environment, the effort to isolate class and race makes very little sense.

Conclusion

"Well, what brings you 'girls' down here?" the sixty-eight-year-old southside activist and retired postal service worker, Manny Herrera, asked when we met for an interview at the Denny's Restaurant near the airport. In other words, why would two Anglo, professional women go to a restaurant that neither had been to before, to meet with a Latino individual whom neither had met before—nor would have met, were it not for this research? Herrera's pointed, and poignant, question sums up an important finding of this research, which is that the Tucson metropolitan area is not a single, homogeneous *community* but rather a segregated city of largely insulated ethnic enclaves. Thus, important units of analysis, when investigating environmental issues in highly populated areas, are the neighborhood and the political district. A trip of a few miles within any large U.S. city can take one into an entirely different culture, and it is *within* these cultures that such social and political constructs as environmental racism either do or do not have validity.

This research further discovered that environmental racism is a product of Pima County's cultural and ethnic heterogeneity. Tucson's Latino community is convinced that it is an accurate description of reality, even if Tucson's white majority is not (or couldn't care less). Moreover, we found that it was not so much the polluting activity itself that brought charges of racism to the fore but, rather, the nature of the official responses to the pollution problem once it was brought into the open.

In our view, this is the most significant finding. In reality, the Anglo majority, the public officials who represented them in the 1980s and 1990s, and mainstream environmental organizations brought this charge on themselves, for they simply did not take seriously the health and safety concerns of people of color and low-income groups. The white majority was too removed, both physically and culturally, from those concerns. "What drives [ordinary] people wild with frustration," the public opinion analyst Daniel Yankelovich wrote recently, "is the lack of responsiveness, a feeling of being ignored, misunderstood, exploited, and played upon like a pack of fools."[28]

Members of Tucson's Latino community felt precisely this way when they raised the TCE issue in 1985. They were told that they had only themselves to blame for their health problems. The subsequent proposal by the county to place a regional landfill in their backyard certainly removed whatever doubts may have remained regarding the reality of environmental racism.

These sentiments grew. Perfectly illustrating the political process model (see Camacho, this volume), by 1994 Latinos had created neighborhood associations, they had elected two younger Latinos to the Board of Supervisors, and there were several prominent southside activists to whom the local media paid attention. They all pulled together to fight the airport dump site.

Their efforts paid off not only in this instance, but in other governmental decisions as well. Thus, for the Latino community, charging the white majority with practicing environmental racism has proven effective. Southside residents were successful in blocking the placement of more detention centers in their neighborhoods, and instead, have attracted a baseball field, a teen center, a new medical complex, and—yes—some golf courses.[29] They also helped force the resignation of the county health director, who was then replaced by a Latino, and they obtained a state-funded health clinic to treat people with TCE-related problems. Despite the seriousness of the Latino community's concerns, however, some Anglo state legislators dismissed the El Pueblo Clinic as simply "pork for Pima County."

Finally, one of Supervisor Grijalva's first actions on taking office in 1989 was to create the Pima County Department of Environmental Quality. The agency began its work by mapping the location of environmental impacts in the Tucson area (see map 2). Because of what the map displayed, the agency's services are often directed into the city's minority areas. That has made the agency controversial, but Grijalva has been an effective defender of both the department and its efforts at cleaning up the city's water and air. For the past five years, for example, the county has tried to enforce stricter emissions regulations than those that are in place for the state as a whole. This has caused officials of Hughes Missile Systems, among others, to object to the higher standards,[30] but Grijalva has stood his ground. In a public exchange in 1994 with Hughes' officials, this time over air quality standards, the supervisor wrote: "Pima County's Department of Environmental Quality is not a renegade department hell-bent on stopping business at any costs, no more than Hughes [is] a callous profit-driven corporation with no civic responsibility."[31]

The decades-long business and governmental practices collectively termed "environmental racism" are changing politics in southern Arizona, for their effects are becoming undeniable. These include higher rates of diseases like lupus and cancer occurring among Tucson's Latino population,[32] stagnant property values for those homeowners living on or near the TCE plume, and air pollution so severe on the San Xavier Reservation that tribal officials recently forced the Minerec Company, which had a lease to operate on their land, to shut down.

Thus, there can be little question that people of color have been placed "at greater environmental risk" than has the Anglo majority in metropolitan Tucson. Although the original decisions to locate industries and other polluting activities in low-income, minority neighborhoods may have had little to do with overt racism, that factor cannot be dismissed in the subsequent history of how the issue was addressed. Indeed, spokespersons for the Latino community never charged "overt racism" in the initial decisions. Rather, as described in the two case studies, it was the fact that Tucson's white majority responded by either ignoring or dismissing the environmental concerns of Latinos. It was only after several years—during which time the Latino community organized, initiated a multimillion dollar lawsuit over water pollution, and elected two Latino supervisors to the county board—that environmental policies in Tucson began to change.

This study highlights two important, and related, features of contemporary environmental politics. First, it shows that people of color and the Anglo majority have quite distinct views on what is important, critical, and worth acting on. "Environmentalism" thus far has not proven to be the consensus-building issue that some thought it would be. To Latinos, it means fighting to keep a garbage dump out of the neighborhood, while for most Anglo environmentalists, it means keeping a hiking trail open in the foothills. In large part, it is the disregard paid to inner-city issues by the white majority, environmentalists included, that has led to charges of racism in Pima County.

The second point is that this research supports the analysis of political representation done by legal scholar Lani Guinier and others: it remains a fact of political life in the United States that people of color generally are best represented by their own members.[33] Even on "common-pool" resource issues, such as clean air and water, issues that logically ought to bring people together, there are significant and enduring differences of opinion among ethnic groups. Race and culture will continue to matter in local policymaking for some time to come.

Notes

A slightly different version of this chapter appears in *Environmental Management* 22, no. 6 (1998) as "Pollution in the Sunbelt," by Jeanne N. Clarke and Andrea K. Gerlak.

1. Bureau of the Census, *April 1990 Census* (Washington, D.C.: Department of Commerce, April 1993).

2. Bureau of the Census, *City-County Data Book,* 12th ed. (Washington, D.C.: Department of Commerce, April 1994).

3. Barry Commoner, *The Closing Circle* (New York: Knopf, 1971); and Commoner, *Making Peace with the Planet* (New York: Pantheon, 1990).

4. Robert D. Bullard, "Race and Environmental Justice in the United States," *Yale Journal of International Law* 18 (1993): 319.

5. We are grateful to Supervisor Raul Grijalva for allowing us to use the various data he has collected. We also appreciate the mapping done by John Regan of the Department of Public Works' Engineering and Technical Services unit.

6. Bill Turque, "In Tucson, 'We Were Victims of War,'" *Newsweek* 6 August 1990, 22. The Tucson case is similar to the Massachusetts one documented by Jonathan Harr in his recent book, *A Civil Action* (New York: Random House, 1995).

7. Pima County Supervisor Raul Grijalva, interview by authors, 21 September 1994.

8. Manuel Herrera, interview by authors, 16 November 1994.

9. Attorney Richard Gonzalez, interview by authors, 7 December 1994.

10. City of Tucson, *Memorandum Regarding Water Pollution Problems—Agency for Toxic Substances and Disease Registry* (Tucson, Ariz.: City Attorney's Office, 23 July 1991).

11. Former Pima County Supervisor David Yetman, interview with authors, 21 November 1994.

12. For a thorough epistemological inquiry, see Michael Polanyi, *Personal Knowledge: Towards a Post-Critical Philosophy* (New York: Harper Torchbooks, 1964).

13. Pima County Supervisor Dan Eckstrom, interview with authors, 26 October 1994.

14. Gonzales interview.

15. Gonzales interview.

16. An epidemiological study of the area just completed by the U.S. Public Health Service has finally validated what the southside residents knew fifteen years ago. See Keith Bagwell, "Report Confirms TCE Health Risk to Southsiders," *Arizona Daily Star*, 15 September 1996, B1.

17. Aaron Wildavsky, "Controversies: Political Culture and Political Preferences," *American Political Science Review* 82, no. 2 (June 1988): 595.

18. Pima County Supervisor Paul Marsh, interview with authors, 16 November 1994.

19. Eckstrom interview.

20. Pima County Supervisor Raul Grijalva, telephone interview, 4 January 1995.

21. Pima County Supervisor Mike Boyd, interview with authors, 26 October 1994.

22. Pima County Supervisor Ed Moore, interview with authors, 2 November 1994.

23. Boyd interview.

24. Eckstrom interview.

25. Grijalva interview, 21 September 1994.

26. Pima County Board of Supervisors, minutes of December 13, 1994 meeting, resolution no. 1994–270. Also, Supervisor Dan Eckstrom, telephone interview with authors, 15 July 1996.

27. Lauretta M. Burke, "Race and Environmental Equity: A Geographic Analysis in Los Angeles," *Geo Info Systems* (October 1993): 44–50.
28. In Jim Hoagland, "U.S. Anger," *Arizona Daily Star,* 10 January 1995, A9.
29. Hippolito R. Corella and Alexa Haussler, "Southside forgotten no longer," *Arizona Daily Star,* 2 June 1996, B1.
30. According to Douglas LaGrange, a major reason for Hughes's objections to the stricter standards is that the company acted *voluntarily* to reduce its volatile organic compound emissions by switching to citrus-based solvents sometime after 1990 (Douglas LaGrange, Pima County Department of Environmental Quality, telephone interview with authors, 1 March 1996).
31. Chris Limberis, "Grijalva attacks Hughes, defends county's environmental agency, director," *Arizona Daily Star,* 22 February 1994, B1.
32. More recent clinical data are confirming higher rates of illnesses in the TCE-contaminated area. See Keith Bagwell, "'High' lupus rate found in TCE clinic's first year 'fit' earlier studies," *Arizona Daily Star,* 9 January 1995, B1.
33. Lani Guinier, "(E)racing Democracy: The Voting Rights Cases," *Harvard Law Review* 108, no. 1 (1994): 109–37.

Kate A. Berry

RACE FOR WATER? Native Americans, Eurocentrism, and Western Water Policy

Discussions about race and racism are rare in writings concerning Native Americans and western water policy.[1] There have, on the other hand, been studies on the availability of clean water for other racial groups, and environmental racism has been discussed concerning Native Americans and nuclear dumping, mining, and fishing.[2] Few would deny that racial tensions exist between Native Americans and Euro-Americans, or that these groups have often competed for water throughout the western United States. Yet little has been done to systematically analyze the connections between these two sets of conflicts.[3]

The emphasis, instead, has been on the legal, economic, and political factors that influence how much and under what conditions Native Americans have access to water. Law has structured the rights and responsibilities of both the federal government and tribal governments, often serving as both the medium and the message of intercultural and intergovernmental communications. As such, tribes and their members have come to depend on law to ensure their right to water. Economics, by contrast, has structured the supply, demand, and market value of resources, including water, for tribes and their members. The economic demand for water revolves around its potential as a basic resource to create and enhance productivity in local, regional, and national marketplaces. It is because water has such a high economic, or commodity, value in the arid West that fierce competition over water rights exists today and that leasing, or marketing, of Native American water has been the subject of much recent interest.[4] Politics, for its part, has mediated the law around and economics of water. Political processes and institutions have reflected cultural norms and the foundation of power more clearly than either law or economics alone.[5] As such, political processes have the potential to reinforce, distort, or even displace the jurisprudence and economic markets that surround Native American water issues.[6]

Even when taken together, however, legal, economic, and political factors cannot fully account for the processes or outcomes of policies that control Native American access to water. Issues related to race and culture have been, and continue to be, instrumental in the allocation of resources. This is because race and culture are ideological in the sense that they provide a foundation for and are imbricated within legal, economic, and political processes.[7] The ideological nature of race and culture refers not to a particular political or partisan association, but rather, to an underlying structural context or set of social dynamics that give race its potency.

Another factor that has hindered analysis is the category of race itself. Using the term *race* in conjunction with Native Americans seems inappropriate to many scholars and, more particularly, to the many Native Americans who do not consider themselves to be a racial minority. The connection is with a particular band, community, tribe, or nation of origin, not with a generalized racial group. One is foremost a Nez Percé, a Hopi, or a Kiowa, not simply a Native American. "American Indians typically think of themselves as members of a particular tribe first and as Indians second."[8] Tribes have rejected racial minority status to assure a strong legal, economic, and political identity as well. "Tribalism remains a driving force in both Indian culture and law. . . . The courts have justified special laws for Indians on the ground that Congress is dealing not with a racial minority but rather with political entities—tribes."[9]

This objection also reflects a broader public rejection of the concept of race as either meaningful or constructive. Race has been considered a vestige of the past that is best left unacknowledged because of its inflammatory potential. Scientific evidence has also been used to deny the importance of race, through the suggestion that physical characteristics and biological factors do not show clear-cut racial categories, thus, race is an irrelevant and unfortunate by-product of cultural conflict.[10]

The impact of race, however, cannot be easily brushed away. While race undoubtedly is an uncomfortable social rather than natural construct, establishing the social nature of the term does not eliminate the significance of race as an idea around which social action and political practices are organized. Nor does the denial of race as a meaningful category eliminate the problems of classification, identity, representation, and recognition that race as a category encompasses.[11] While reducing race to an autonomous component is inherently problematic, the idea of race, nevertheless, has become institutionalized and has a potent force. Race has become an important category due to the distribution of power in society, that is, because there are power differentials based on race. To ignore race is, in effect, to ignore racism, as well as the inequities or denial of access to power and resources that it implies.[12] In essence, racism is a set of postulates, images, and practices that serve to differentiate and dominate on the basis of race.[13]

The broad and ugly web of racism does not unravel willingly and is certainly a task beyond the scope of this chapter. The more modest scope here is to examine a particularly tenacious formulation of racism, Eurocentrism, as a determinant in public water policies that influence Native Americans.

Commonly, racism and Eurocentrism are looked on as distinctly separate entities. Racism is confined to situations in which people are identified as superior or inferior based on inherited biological grounds, and Eurocentrism is limited to situations in which people are identified as superior or inferior based on cultural grounds. By contrast, the approach adopted here collapses these categories. Based on the ample evidence that biology alone cannot be determinative of race and, hence, used to support racism, Eurocentrism is considered to be a formulation of racist ideology, a manifestation of a specific variant of racism. While Eurocentrism cannot simply be equated with racism, it is a key strand in the ideological web of racism.

Eurocentrism is based on the notion that European civilization "has had some unique historical advantage, some special quality of race or culture or environment or mind or spirit, which gives this human community a permanent superiority over all other communities, at all times in history and down to the present."[14] As a particularly powerful and ingrained belief system, Eurocentrism implies not only the privilege of European norms, values, institutions, and peoples, but also the active and conscious diffusion of this ideology outward from a center in Europe.[15] Of special interest to this chapter is the legitimizing, energizing, and constraining role of

Eurocentrism that is embedded in legal, economic, and political discourse: "For half a millennium . . . Western legal thought has sought to erase the difference presented by the American Indian in order to sustain the privileges of power it accords to Western [European] norms and value structures."[16]

The argument developed here is that Eurocentrism has been instrumental in the development, evolution, and outcome of policies that affect the access of Native American people and tribes to water. This is not to suggest that Eurocentrism is monolithic as the determinative influence on Native American water policy. Rather, this approach proposes that an enhanced understanding of the nature and extent of Eurocentrism is critical to decipher the historical and geographical positioning of policy processes and outcomes.

This chapter considers both how Eurocentrism impacts water policymaking processes and institutions as well as how these processes and institutions reinforce Eurocentrism. The objective is to get a sense of the broad impact of Eurocentrism by examining several situations at various times and places within the western United States:

> Indian water rights, as they relate to western water development, must be seen against the backdrop of the history of the hemisphere. Ever since the arrival of the Europeans on the continent, an important current in the development of the legal system has been to define Indian rights and then develop an orderly process for taking them away. From the formulation of the doctrine of discovery itself, this two-step exercise has served the humanitarian purpose of attempting to accord some fairness to Indians while providing discipline to the competition among non-Indians for the right to use Indian resources.[17]

The Colonial Spanish Legacy

Even before setting foot in the Americas, Spain had institutionalized its vision of discovery and conquest. Codified in 1265, *Las Siete Partidas* set forth a mandate for human dominion over nature and over one another: "Man has power to do as he sees fit with those things that belong to him according to the laws of God and man."[18] Early invasions of Spain by the Arabs, Romans, Germans, and

Moors left their mark on the Spanish legal/political system, and extended a notion of conquest and social stratification into many aspects of Spanish law, including their policies toward the water and land rights of the Native American cultures that the Spanish encountered.

Once situated in the Americas, Spanish law evolved to meet the different conditions, but it maintained as an intrinsic characteristic the hierarchy of public ownership and individual property rights. The underlying premise in New Spain was that the Spanish Crown owned everything it "discovered" or conquered. Thus all property titles, including land and water, stemmed from the Crown because everything belonged to the sovereign as discoverer.

New World Spanish law, *Los Leyes de los Indios,* included a provision that water was to be used for the "common good." The Spanish, however, were the ones to interpret the meaning of the common good; their determination of what constituted the best use of water was confined to Eurocentric, rather than Native American, notions of appropriate allocation and management.[19]

Each Native American culture or tribe, however, had experience using and adapting to limited water supplies that existed prior to their initial encounter with the Spanish and other Europeans. Some groups—for example, the native people of the upper Rio Grande basin, today collectively referred to as the Pueblo Indians—had extensive experience with small-scale irrigated agriculture. Different strategies of irrigation evolved to sustain corn, cotton, wheat, melons, pumpkins, onions, beans, and chilies. Zuni Pueblo women, for instance, transported water from streams in jars on their heads. In Acoma Pueblo, by contrast, natural depressions in the rocks that received rainfall runoff from the upper mesas were utilized. Many Pueblo groups used stream diversions or built small storage reservoirs to irrigate their fields. Water also took on an importance to the early Pueblo people beyond its value for irrigation. It was seen as part of a cyclical balance between the earth and sky that should not be undermined.[20]

Further south and west, the desert O'otam, or Papago Indians, had a somewhat different relationship with water. As hunters and gatherers, the O'otam lived in small clans and migrated in search of water sources. Access to water meant survival, but it was also fundamental to their worldview:

Papago life was premised on the fact that water was constantly in motion and that humans could endure only by fitting into this movement. The lesson was written across the landscape of the Sonoran desert. Vegetation clung to the arroyos and ephemeral rivers. As one moved off these topographical concentrations of water, life thinned with the dilution of moisture. A man could walk from alpine forest to cactus bajada in a day and witness a succession of floral worlds separated by gradations of rainfall. The teachings of the land left their mark. The language of the people was saturated with words derived from one thought: water. The great rituals of the O'otam focused on songs to pull down the clouds.[21]

In most instances, the Spanish did not recognize or support the prior water use and practices of the Native Americans that they displaced. Instead, the Spanish instituted a system of land and water grants made by the Crown, or its designate, to those whom it favored. This system created extensive property holdings called *haciendas,* which were subsequently consolidated through purchase, plunder, inheritance, marriage, and donation. Control and management of water was linked to its potential for political and economic dominance because water affected the productive capacity of the land.[22]

In a few instances, such as some of the Pueblo Indians in northern New Mexico, the Spanish recognized cultures as "civilized" and, hence, they were deemed to "own" the real property they possessed at the time they were "discovered." But European norms prevailed as the power relationship between the "conqueror" and the "conquered" remained basically intact, even in the case of the Pueblo Indians. The conquering Spanish subjected the Pueblo Indians to Spanish law and considered them to be wards of the benevolent Crown. As such, the tribes were given grants to land and water, and in turn, the Crown was to protect them from encroachment on their land and resources. Inevitably, however, encroachment did occur as subsequent land grants made to Spaniards overlapped with land and water granted to the Pueblo Indians.[23]

In this less obtrusive expression of Eurocentrism, Native Americans were treated akin to children who did not have the capacity, skills, or experience to understand the complexity of Spanish thought and civilization. This paternalistic Eurocentrism is evident in the preferential treatment given to Native Americans in some situations, such as the transfer of property, both land and water:

Not only did Spain enact ordinances of special stringency to protect Indian lands against trespass, but it sought likewise to protect Indians against the superior bargaining power of the white race by outlawing all transfers of Indian property not made before an appropriate judicial officer under conditions of notice designed to bring to the Indian an adequate return for that which he sold.[24]

In a more conspicuous form, Eurocentrism was evident in the utilization of Native Americans as a means to achieve the legal, economic, or political ends desired by Spaniards. An institutionalized example of this type of exploitive Eurocentrism was *encomiendas,* or Crown grants to Native Americans who labored for Spaniards who were favored. Despite a protracted moral and legal debate in Spain over the status and rights of Native Americans, in practice, Native American labor became virtually indistinguishable from slavery under the system of *encomiendas.* Landless and waterless Native Americans were given access to their original resources; in exchange, the entire family worked for the *hacienda,* with the obligation often enforced for generations. The outcome of this system of land and labor was the availability of virtually free labor for agricultural activities and domestic service for the *hacienda;* moreover, productive processes were organized in such a way that surpluses were concentrated almost exclusively in the hands of Spanish *hacienda* owners and their intermediaries.[25]

Spanish law has had a longstanding impact throughout the area that is now the southwestern United States. Mexico, after achieving independence in 1821, adopted and adapted many of the basic tenets of the Spanish law—although important changes were made. The institutions of land grants, *haciendas,* and *encomiendas* were not eliminated outright, though certain aspects of Native American wardship status were changed. Under Mexican rule, Native Americans could own land and sales of their lands were permitted in order to redistribute unused communal lands.[26] Mexican water law, however, restated the Spanish presumption that streams belonged to the nation and that the use of water belonged to the inhabitants in common.

Although the United States took possession of the lands in the southwest one and a half centuries ago under the treaty of Guadalupe Hidalgo, the Spanish past continues to have contemporary ramifications. Questions persist about the presumptions, responsibilities, and actions of the colonial Spanish state on the allocation of water for Indian tribes. Current judicial interpretations of colonial land titles, associated water rights, and agricultural patterns continue to influence the dimensions of water rights for Native American tribes, despite the distinctly different legal doctrines that govern water allocation today.[27] As questions about colonial Spanish tradition and law wind their way through the courts, old Spanish norms are reinterpreted in light of more current Eurocentric notions regarding the appropriate distribution of water for Native American tribes.

U.S. Government Policy

There has always been a rather complex relationship between the U.S. government, created principally for and by Euro-Americans, and Native American tribes. Native American affairs have traditionally been considered concerns of the nation and, as such, the federal government has long enjoyed a virtual monopoly in its dealings with tribes. Following in the footsteps of eighteenth-century British treaty makers, the United States early on assumed plenary power as domestic sovereigns in dealing with Native American tribes. This relationship has not been between two equals, but rather, "their [Native Americans] relation to the United States resembles that of a ward to his guardian."[28] The trust responsibility of the federal government derives from the original treaties and agreements between federal and tribal governments. Federal policy and law, in large measure, is a derivation of this federal trust responsibility to the tribes.

This relationship, based on the roles of guardian and ward, evolved from and was fostered by paternalistic Eurocentrism on the part of the federal government and an associated dependency on the part of Native Americans. Paternalistic Eurocentrism is evident in three fundamental principles that shaped the creation of federal policy. The first principle was that all humankind was one and, therefore, that all human beings were created innately equal. The second was that Native Americans in their existing cultural circumstances were inferior to Europeans and Euro-Americans. And the third, that Native American culture and people could and should be transformed to equal or approximate that of their Euro-American neighbors.[29]

Similar to the benevolent, yet decidedly Eurocentric, intent of some of the Spaniards, the actions of many humanitarian reformers in the United States were based on paternalistic but well-meaning intentions, as well as the naive belief that with guidance and protection Native Americans would move quickly toward Euro-American beliefs and practices. The darker side of Eurocentrism, with an explicitly exploitive intent, was manifest in the interests of Euro-Americans being served at the expense of Native American rights and well-being.

Federal policy also contains unresolved tensions between two conflicting objectives—assimilation and separation—that are rooted in the nature and expression of Eurocentrism. Assimilation, the idea of incorporating Native Americans into the Euro-American melting pot, has at its core the Eurocentric assumption that Euro-American culture, law, economics, and politics are superior to their Native American counterparts. As a result, assimilation has been a recurring, if episodic, objective of congressional policy and has had many congressional champions over the years, including Senator James Warren Nye of Nevada, who in 1871, said:

> We see nothing about Indian nationality or Indian civilization which should make its preservation a matter of so much anxiety to the Congress or the people of the United States. The fundamental idea upon which our cosmopolitan republic rests is opposed to the encouragement or perpetuation of distinctive national characteristics and sentiments in our midst. We see no reason why the Indian should constitute an exception. . . . If the Indian cannot learn to forego such of habits as are peculiar to savage life, and such of his political opinions and sentiments as are not in harmony with the general policy of our Government, then he cannot, beyond a limited period exist among us, either as a nation or as an individual.[30]

By contrast, the federal policy objective of separation aims to form separate social and physical spaces for Native Americans through the creation of the reservation system. The goal of separation encompasses both variants of Eurocentrism, paternalistic and exploitive. As Francis Paul Prucha develops the logic, "If the Indians came to rely on agriculture and domestic manufacture for their food and clothing, they would no longer need extensive hunting grounds and would willingly give up their unneeded lands for white settlement."[31]

The role of federal policy in contemporary Native American life is enormous. The legal federal–Native American trust relationship structures everything from the role of tribal governments to the health care, education, and social services available to tribes and individual members. Many Native American rights, including water rights, have been supported and defended on the basis of the federal trust responsibilities, treaties, and subsequent laws that have evolved from early treaties and agreements. The episodic vacillation of federal policy objectives from assimilation to separation further undermined the foundation for Native Americans to defend their rights and resources.

Native American Water Policy and Reserved Water Rights

In the historical chronicles of western water policy, the origins of the prior appropriation doctrine are usually traced to nineteenth-century Mormon pioneers who protected the use of water by early settlers against any incursion of later settlers and to early mining laws, which also protected first use.[32] Many arid western states adopted the standard of first use to prioritize and allocate water among competing users. Little regard was paid to the preexisting water uses and interests of Native Americans. Tribes did not typically have the capital, infrastructure, incentive, or resources to appropriate water without relying on the federal government.

Congress took little initiative to identify the basis for Native American water rights, in dramatic contrast to the formulation of other types of federal policy. While Congress has acted to authorize and appropriate funds for individual water projects for specific tribes, it has not legislated policy on whether or why or by what means Native Americans have rights to water. The vacuum left by Congress eventually moved Native American water policy into the judicial arena.[33]

The landmark judicial action involving Native American water rights was the 1908 U.S. Supreme Court case, *Winters v United States.*[34] This case involved a conflict over the waters of the Milk

River in northern Montana between the Native Americans of the Fort Belknap Reservation and Euro-Americans who had settled in the valley after the tribal land base was diminished in 1888. In 1905, the Bureau of Indian Affairs (BIA), which wanted to develop irrigation on the reservation, requested that the Department of Justice take action against the settlers to ensure an adequate supply of water for Native American crops.

In its capacity as trustee for the Indians, the Department of Justice filed suit in a district court based on several arguments. First, as part of the Fort Belknap Reservation, the Milk River was never a source of public water subject to appropriation. Second, the reservation held water rights in sufficient quantity to carry out the purposes for which the reservation was created. Third, state legislation cannot destroy the water rights of the federal government. Finally, depriving Native Americans of water would violate the treaties and agreements between the Native Americans and the United States government. Many federal policymakers and states' rights' advocates were incensed with this approach. It was a break with tradition to claim federal water rights in the western United States; in concurrent cases, the federal government stood firmly committed to the first use appropriation system.[35]

The Department of Justice won the case for the Fort Belknap Reservation in the district court, then in the Ninth Circuit Court of Appeals, and again in the Supreme Court. The Supreme Court decision forms the foundation of the reserved right, or *Winters* doctrine. While the opinion was brief and left many unanswered questions, the Supreme Court clearly upheld the tribe's superior rights to the waters of the Milk River.

Through the *Winters* and subsequent court decisions, reserved rights were distinguished from traditional state water law and the doctrine was fleshed out. Native American reserved water rights are founded on the original treaties and agreements between the federal and tribal governments, which reserved not only land, but implied a reservation of water to make the land usable. Thus reserved water rights were designed to accommodate the purpose(s) of the reservation, as set forth in congressional treaties, acts, or executive orders, with the right and priority to use water typically based on the date when the reservation was created. Unlike the "use it or lose it" philosophy of the prior appropriation doctrine, *Winters* rights are reserved indefinitely and are not subject to any test of their beneficial use.

The holding of the *Winters* case was unexpected. Up to that point, only two doctrines of water allocation had been enunciated: prior appropriation and riparian. Both were based on state, rather than federal, law and were supported by legislative statutes. The perspective of existing western water users was that the imposition of a new and contradictory doctrine only complicated matters, making the process of allocating water less secure. They expected Congress to continue to defer to state water law. Even before the Supreme Court holding was made public, some water users organized into interest groups and began lobbying Congress for relief.[36]

During the same period, the BIA and several eastern Congresspersons spearheaded an effort to legislatively enact the *Winters* doctrine. The BIA apparently did not think that the court decision would be implemented without congressional approval. The dispute between proponents and opponents of the *Winters* doctrine resulted in debates over various bill proposals and left Congress at an impasse. According to Daniel McCool:

> There were plenty of opportunities to legislate the Winters doctrine out of existence or officially recognize it. Congress did neither. It appears that both sides in the conflict had sufficient political clout to exercise a veto effect on such legislation but insufficient strength to overwhelm the opposition.[37]

In the decades since the *Winters* decision, Congress has periodically revisited the issue, but it has been unable to develop a coherent water policy for Native American tribes. As a result, Congress authorizes and appropriates funds for Native American water development on an ad hoc basis and only infrequently sanctions specific water rights for tribes. The bureau's program of developing Native American water resources remained virtually unchanged: "In short the BIA chose to minimize the *Winters* doctrine and the impact it might have on Indian water rights and . . . generally bowed to the political necessity of following state water law."[38] The *Winters* doctrine was resorted to only as a last measure.

From 1908, when it was first formulated, until its reaffirmation by the Supreme Court in the 1960s, the *Winters* doctrine has remained relatively obscure and nonthreatening. In 1973, the National Water Commission observed that the right of Native Americans to water was not dealt with equitably relative to the right of non–Native Americans:

> Following *Winters*, more than 50 years elapsed before the Supreme Court again discussed significant aspects of Indian water rights. During most of this 50 year period, the United States was pursuing a policy of encouraging the settlement of the West and the creation of family-sized farms on its arid lands. In retrospect, it can be seen that this policy was pursued with little or no regard for Indian water rights and the *Winters* doctrine. With the encouragement, or at least the cooperation, of the Secretary of the Interior many large irrigation projects were planned and built by the Federal Government without any attempt to define, let alone protect, prior rights that Indian tribes might have had in the water used for the projects. . . . In the history of the United States Government's treatment of Indian tribes, its failure to protect Indian water rights for use on the Reservations it set aside for them is one of the sorrier chapters.[39]

Reserved water rights have provided an important legal basis for Native Americans to claim, and actually receive, water in the past three decades. The failure to adequately implement the policy for over three-quarters of a century, however, represents the precedence of Eurocentric rights, norms, and values over Native American rights, norms, and values. An interest in promoting agrarian, municipal, and industrial development, designed by and for Euro-Americans, was deemed to be more important than ensuring adequate water for reservations.

The Pick-Sloan Water Development Plan

Water law and water rights have not proven to be the only areas where Eurocentric values and interests have shortchanged Native American tribes. The most tangible, and perhaps most devastating, impacts on tribes have resulted from the design, construction, and operation of water development projects. In attempting to manage water as a natural resource, water projects are designed to control floods, improve navigation, generate power, and secure water supplies.[40] Yet the toll in human, as well as ecological, terms has been enormous and Native Americans on reservations have shouldered these costs disproportionately, realizing few of the benefits.

Among the most pernicious projects is Pick-Sloan, a massive water development plan for the upper Missouri River basin developed jointly by the U.S. Bureau of Reclamation and the U.S. Army Corps of Engineers in 1944. The Pick-Sloan plan ultimately affected twenty-three different reservations throughout Montana, Wyoming, Nebraska, North Dakota, and South Dakota, and is said to have caused more damage to Native American lands than any other public works project in the United States. Native American homes, ranches, and communities were inundated, forcing more than 900 families to relocate. Reservations became cut off from basic services, schools, and health care facilities, and many were isolated from communication and transportation networks. In North and South Dakota alone, five mainstem water projects reduced the land base of reservations by over 550 square miles. Among the hardest hit by the extensive Pick-Sloan plan were six reservations in the Dakotas: Fort Berthold, Standing Rock, Cheyenne River, Yankton, Crow Creek, and Lower Brule.[41]

The most severe impact to tribal life from the Pick-Sloan plan was to the Fort Berthold Reservation, home to the Mandan, Arikara, and Hidatsa peoples—the Three Affiliated Tribes. Many decades before water development projects were contemplated, members of the Three Affiliated Tribes were exposed to Eurocentrism. Early European and Euro-American travelers had a particular fascination with the relatively fair skin coloring and presence of blue eyes among the Mandan people. Unable to accept the Mandon's New World ancestry, some travelers from the Old World concluded that they were the descendants of Europeans. As Joseph Cash and Gerald Wolf explained, several theories evolved on the European origins of the Mandan people:

> There was one theory that the Mandan were part of the ten lost tribes of Israel, and yet another that claimed they were the descendants of Irish monks. In particular, George Catlin advanced the notion that they were descendants of the Welsh, and the famed artist spent much of his life trying to trace similarities between the Mandan and Welsh languages. He was convinced that his idea was correct.[42]

For nearly 700 years, the Mandan, Arikara, and Hidatsa peoples were supported by the deep wooded bottomlands and benchlands along the upper Missouri River. Although life was not idyllic, forage, food and medicinal plants, fish, game, water, and shelter were found in this fertile area. They grew a variety of crops, including potatoes, beans, squash, several types of corn, and other vegetables. Lignite for fuel and timber for building were available on the adjacent hillsides.[43]

The lifestyle and culture of the Three Affiliated Tribes were severely disrupted by the Garrison Dam and Lake Sakakaewea, one of the earliest mainstem dams and reservoirs of the Pick-Sloan plan. The corps flooded over one-fourth of the reservation land base, 152,360 acres, including all the fertile river bottomlands and benchlands and 94 percent of the agricultural lands. Without access to fertile river bottomlands, the residual lands of the reservation were unproductive. The devastation to these tribes was extensive and included: the relocation of 80 to 90 percent of the tribal membership; the relocation of the agency headquarters and boarding school; the loss of seven day schools on the reservation; the loss of 80 percent of the reservation's road system; the loss of coal mines, timber, fruit, and berries; and the denial of fishing, grazing, and salvage logging privileges as well as mineral rights. The water project also physically divided the reservation into five isolated districts. Further fragmentation occurred as kinship groups that had lived in close proximity for generations were scattered.

In addition to the high costs, Fort Berthold residents realized few of the Garrison Dam's benefits. A petition for a block of the Garrison project's power was denied, and access to irrigation water and water rights still has not been reconciled.[44] After many years of dispute in the halls of Congress, the courts, and administrative committees, federal legislation was passed in 1991 to provide compensation for the taking of land and resources from the Three Affiliated Tribes. An economic trust fund was established to provide interest income to the tribes and their members, and 1,937 acres of nontribal lands were transferred to the tribes.[45]

The five Sioux tribes of the Standing Rock, Cheyenne River, Yankton, Crow Creek, and Lower Brule Reservations also sustained significant damages from Pick-Sloan, although not of the same magnitude as the Fort Berthold tribes. Dams on the Missouri River inundated more than 200,000 acres on these five reservations, reducing their land base by about 6 percent. The impact, however, was far greater than this figure might suggest because each of the reservations' population was concentrated on lands near the river.

Over one-third of the population of the five reservations were forced to relocate permanently, some in the dead of winter, others without adequate funds to replace their former homes. Their new homes in more marginal prairie lands could not replace the ready access to fish, game, plant medicine, food gathering, livestock forage, shade, arable land, fuel, and water in the fertile river bottom lands that they had been forced to abandon.[46] The relocations prompted significant changes in lifestyle and caused great hardship, as Michael Lawson described:

> Approximately 580 families were uprooted and forced to move from rich sheltered bottomlands to empty prairies. Their best homesites, their finest pastures, croplands and hay meadow, and most of their valuable timber, wildlife and vegetation were flooded. Relocation of the agency headquarters on the Cheyenne River, Lower Brule and Crow Creek reservations seriously disrupted governmental, medical, and educational services and facilities and dismantled the largest Indian communities on these reservations. Removal of churches, community centers, cemeteries, and shrines impaired social and religious life on all five reserves. Loss not only of primary fuel, food, and water resources but also of prime grazing land effectively destroyed the Indians' economic base. The thought of having to give up their ancestral land, to which they were so closely wedded, caused severe psychological stress. The result was extreme confusion and hardship for tribal members.[47]

As with the Fort Berthold tribes, the five Sioux reservations have not shared in the rewards of these water projects. They have not had access to inexpensive electricity from hydropower; indeed, extensive areas within the reservations are still without electric service because it's too expensive. While the Standing Rock Sioux tribe recently received compensation through the establishment of an economic development trust fund and funding for irrigation development, the other tribes have never had the access to water or the capital needed for irrigated agriculture nor adequate municipal and domestic water supplies.[48] And even in recreation and tourism, the most successful aspect of the Pick-Sloan projects, Indians have reaped only limited benefits due to their lack of venture capital as well as the pervasive anti–Native American biases in the Euro-American communities of South Dakota.[49] The drive for more judicious compensation continues to be a pressing issue for these tribes and others throughout the Missouri River basin. In the 1990s, a coalition of twenty-one tribes is developing the expertise necessary to rectify past problems and prevent future encroachments by the Bureau of Reclamation and the Army Corps of Engineers.[50]

Perhaps the most culpable aspect of the entire Pick-Sloan plan was the Eurocentric nature of the planning and compensation processes. The water projects were planned and designed to benefit the Euro-American community, rather than the Native American tribes of the upper Missouri basin. Although they sustained a disproportionate burden, none of these tribes were consulted in the planning or development of the water project.[51] "The Pick-Sloan Plan was thus presented to the Sioux as a fait accompli. The federal government was determined to move them out, and there was nothing that they could do about it," observed Lawson.[52]

The process used to compensate tribes and tribal members for their lands and the hardships of relocation was equally dismal. Each tribal government negotiated independently and under different circumstances, which led to disparities between the various tribes in the terms and amounts of compensation. In all cases, however, the tribes were at a disadvantage as negotiations took place only after the plan had been authorized by Congress and monies had actually been spent on project planning and construction. The corps also flexed its negotiating muscles in the Fort Berthold, Yankton, Crow Creek, and Lower Brule tribal negotiations by condemning tribal lands under the power of eminent domain without regard to the long-standing federal trust relationship that required Congress to act on any takings of reservation lands.[53]

In 1962, President Kennedy formally dedicated one key component of the Pick-Sloan plan, the Oahe Dam, proclaiming: "This dam provides a striking illustration of how a free society can make the most of its God-given resources."[54] Encapsulating the prevailing Euro-American imagery of water project development, this statement is evidence of the depth of Eurocentrism that surrounded the Pick-Sloan plan. The quest for flood control, improved navigation, and to reclaim arable lands in the upper Missouri basin was propelled by the desire to improve the condition of Euro-American communities. These communities were largely spared displacement and hardships, and were the targets of the anticipated benefits.

The rhetorical approach taken by President Kennedy illustrates the potency of European-derived values and norms in water resource development. Eurocentrism, in its exploitive expression, is quite explicit in the link made between resources and independence. This expression of Eurocentrism is also apparent more broadly throughout the course of the planning and implementation of the project. A more paternalistic expression of Eurocentrism was manifest in the subsequent reparation payments made to the Three Affiliated Tribes, although this action, occurring long after construction, could not diminish the severity of impact to the six reservations. The myopic vision that supported the management and development of Missouri basin water was incapable of fully considering the values, norms, and conditions of the Native American people of the region.

Pick-Sloan is not the only water development plan that represented Eurocentric values at the expense of native cultures, values, and rights. The plight of these six reservations has reverberated throughout many other native communities and reservations in the United States—such as the Crow and Salish Kootenai communities in Montana—as well in native communities in other parts of the world—such as the Cree in northern Canada, the Penga and Dom in India, the Akawaio in Guyana, and the Purari in New Guinea. Some, such as the Bontoc and Kalinga peoples of the Philippines, are actively resisting the encroachments of water development projects on their cultures, lifestyles, and lands, and have met with fierce opposition and oppression by government.[55]

Conclusion

I have chosen several different situations to establish the point that Eurocentric notions are deeply embedded within the legal, economic, and political processes and institutions that distribute water in the United States. Eurocentrism is not limited to selectively chosen government officials who act against the best interests of Native Americans because of an atypical racial bias. Instead, these examples have been presented as indicators of the enormous effect that European standards and norms have had on water rights and water development institutions that impact Native Americans.

It is not necessary or possible to attempt to distinguish between policies that are based on Eurocentrism and others that are not influenced. From the initial European conquest of the Americas to the present, Eurocentrism has been widely accepted and pervasive; it is simply a question of degree and character.

There has been slight recognition, much less accommodation, of alternate Native American uses, values, and processes for distributing water. Instead, the accommodation has largely been in the other direction—Native Americans have had to accommodate themselves to water distribution systems and policies of European origin. In some cases, accommodation has been successfully negotiated, but in many instances, the result has been disastrous. The experience of the Three Affiliated Tribes with the Garrison Dam project is a potent example. As Roy Meyer explained: "From the [Mandan, Arikara, and Hidatsa] Indian point of view, the building of the dam and its consequent destruction of the old way of life constituted yet another example of the white man's persistent effort to force the native people of this continent to become like himself."[56]

While it may be relatively simple to label a person, an action, or even the outcome of a policy as racist, it is much more difficult, but arguably more important, to probe the recurring patterns and processes that bind racial and cultural identity and norms, such as Eurocentrism, to the realms of culture, law, economics, and politics. At one level, the conclusion that western water policy is Eurocentric seems apparent, yet the ramifications and significance of this have mostly been denied or ignored by many who participate in the policy process.

This chapter will be successful to the degree that it expands the dialogue to incorporate the topic of Eurocentrism in Native American water issues. The intent has been to reveal the foundations on which individuals and institutions act and policies evolve on the basis of race and cultural norms. The challenge is not simply to catalog abuses but to identify and take action to rectify the policy patterns and processes that support inequities in the distribution of water based on Eurocentrism.

Notes

Judith Jacobsen, Daniel McCool, Laura Pulido, and Elmer Rusco provided insightful comments on an early draft of this chapter.

1. An exception to this can be found in Daniel McCool, *Command of*

the Waters: Iron Triangles, Federal Water Development, and Indian Water (Berkeley: University of California Press, 1987), 157–58.

2. See the chapters, this volume, by White; Bath, Tanski, and Villarreal; and Robyn and Camacho. See also Patrick C. West, J. Mark Fly, Frances Larkin, and Robert W. Marans, "Minority Anglers and Toxic Fish Consumption: Evidence from a Statewide Survey of Michigan," in *Race and the Incidence of Environmental Hazards: A Time for Discourse,* ed. Bunyan Bryant and Paul Mohai (Boulder, Colo.: Westview Press, 1992), 100–113; and Gail Small, "Environmental Justice in Indian Country," *Amicus Journal* (spring 1994): 38–40.

3. This is not to suggest that all water rights conflicts are between Euro-Americans and Native Americans. There are water conflicts between Hispanics and Native Americans, one tribe and another tribe, pro-development Native Americans and traditional Native Americans, etc. Rather, the argument here is that the role of race and racism have not been fully explored in analyzing conflicts between Native Americans and Euro-Americans over water.

4. John Folk-Williams, *What Indian Water Means to the West* (Santa Fe, N.Mex.: Western Network, 1982); David H. Getches, "Management and Marketing of Indian Water: From Conflict to Pragmatism," *University of Colorado Law Review* 58, no. 4 (1988): 515–49; Rodney T. Smith, "Water Rights Claims in Indian Country: From Legal Theory to Economic Reality," in *Property Rights and Indian Economies,* ed. Terry L. Anderson (Lanham, Md.: Rowman and Littlefield, 1992), 167–94.

5. McCool, *Command of the Waters;* and Donald Worster, *Rivers of Empire: Water Aridity and the Growth of the American West* (New York: Pantheon, 1985).

6. With the convergence of law, economics, and politics have emerged pragmatic actions such as negotiated water rights settlements. For more on negotiated settlements on Indian water rights, see John E. Thorson, "Resolving Conflicts through Intergovernmental Agreements: The Pros and Cons of Negotiated Settlements," in *Indian Water 1985,* ed. Christine Miklas and Steven Shupe (Oakland, Calif.: American Indian Resources Institute, 1986), 25–47; Lloyd Burton, *American Indian Water Rights and the Limits of Law* (Lawrence: University of Kansas Press, 1991), 63–86; Lois Witte, "Negotiating an Indian Water Rights Settlement: The Colorado Ute Indian Experience," in *Innovation in Western Water Law and Management,* proceedings of the 12th annual summer program of the Natural Resources Law Center, University of Colorado School of Law, 5–7 June 1991; Daniel McCool, "Intergovernmental Conflict and Indian Water Rights: An Assessment of Negotiated Settlements," *Publius* 23, no.

1 (1993): 85–101; and Norman H. Starler and Kenneth G. Maxey, "Equity, Liability, and the Salt River Settlement," in *Indian Water in the New West,* ed. Thomas R. McGuire, William B. Lord, and Mary G. Wallace (Tucson: University of Arizona Press, 1993), 125–46.

7. For more on the need to incorporate the ideological nature of race into environmental racism research, see Laura Pulido, "A Critical Review of the Methodology of Environmental Racism Research," *Antipode* 28, no. 2 (1996): 142–59.

8. David H. Getches and Charles F. Wilkinson, *Cases and Materials on Federal Indian Law,* 2d ed. (St. Paul, Minn.: West Publishing, 1986), 1.

9. Ibid.

10. For a discussion of why racial categories are cultural constructs masquerading as biology, see Jonathan Marks, "Black, White, Other," *Natural History* (December 1994): 32.

11. Kay J. Anderson, "The Idea of Chinatown: The Power of Place and Institutional Practice in the Making of a Racial Category," *Annals of the Association of American Geographers* 77, no. 4 (1987): 581. See also Itabari Njeri, "Sushi and Grits: Ethnic Identity and Conflict in a Newly Multicultural America," in *Lure and Loathing: Essays on Race, Identity, and the Ambivalence of Assimilation,* ed. Gerald Early (New York: Penguin, 1993), 37–40.

12. Adil Qureshi, h-net ethnic history discussion internet list, 7 December 1994.

13. Floya Anthias and Nira Yuval-Davis, *Racialized Boundaries* (London: Routledge, 1993), 15.

14. James M. Blaut, *The Colonizers Model of the World: Geographical Diffusionism and Eurocentric History* (New York: Guilford Press, 1993), 1.

15. Ibid., 1–43. See also Calvin Martin, *The American Indian and the Problem of History* (New York: Oxford University Press, 1987); and Teun A. van Dijk, *Elite Discourse and Racism* (Newbury Park, Calif.: Sage, 1993).

16. Robert A. Williams Jr., *The American Indian in Western Legal Thought: The Discourses of Conquest* (New York: Oxford University Press, 1990), 326.

17. Sam Deloria, "A Native American View of Western Water Development," *Water Resources Research* 21, no. 11 (1985): 1785.

18. Quoted in Norris Hundley Jr., *The Great Thirst: Californians and Water, 1770s–1990s* (Berkeley: University of California Press, 1992).

19. William W. Robinson, *Land in California* (Berkeley: University of California Press, 1948). See also Williams, *American Indian in Western Legal Thought.*

20. Michael C. Meyer, *Water in the Hispanic Southwest: A Social and Legal History, 1550–1850* (Tucson: University of Arizona Press, 1984), 15–23.

21. Charles Bowden, *Killing the Hidden Waters: The Slow Destruction of Water Resources in the American Southwest* (Austin: University of Texas Press, 1977), 8–9. For more on comparative ideologies of water in the context of two Paiute-Shoshone communities and a Euro-American ranching community, see Kate A. Berry, "Of Blood and Water," *Journal of the Southwest* 39 (1997): 79–111.

22. Meyer, *Water in the Hispanic Southwest,* 21–42.

23. Malcolm Ebright, "New Mexican Land Grants: Their Legal Background," in *Land, Water, and Culture,* ed. Charles L. Briggs and John Van Ness (Albuquerque: University of New Mexico Press, 1987); and G. Emlen Hall, "The Pueblo Grant Labyrinth," in *Land, Water, and Culture,* ed. Charles L. Briggs and John Van Ness (Albuquerque: University of New Mexico Press, 1987).

24. Felix Cohen, "The Spanish Origin of Indian Rights in the Law of the United States," *Georgetown Law Journal* 31 (1942): 15, originally cited in Daniel McCool, "Precedent for the Winters Doctrine: Seven Legal Principles," *Journal of the Southwest* 29, no. 2 (1987): 171.

25. Malcolm Ebright, *Spanish and Mexican Land Grants and the Law* (New York: Sunflower University Press, 1989).

26. Cecil Alan Hutchinson, *Frontier Settlement in Mexican California* (Berkeley: University of California Press, 1969).

27. Frances Levine, "Dividing the Water: The Impact of Water Rights Adjudication on New Mexican Communities," *Journal of the Southwest* 32, no. 3 (1990): 268–77; see also Hall, "Pueblo Grant Labyrinth."

28. *Cherokee Nation v Georgia,* 30 US (5 Pet) 1.

29. These principles have been adapted from a discussion of paternalism and its repercussions in Indian country by Francis Paul Prucha, *The Indians in American Society: From the Revolutionary War to the Present* (Berkeley: University of California Press, 1985), 8–10.

30. Quoted in Sharon O'Brien, *American Indian Tribal Governments* (Norman: University of Oklahoma Press, 1989), 71.

31. Prucha, *Indians in American Society,* 12.

32. Ned K. Johnson and Charles T. DuMars, "A Survey of the Evolution of Western Water in Response to Changing Economic and Public Interest Demands," *Natural Resources Journal* 29, no. 2 (1989): 347–87.

33. McCool, *Command of the Waters,* 61–65.

34. *Winters v the United States,* 207 US 564 (1908).

35. McCool, *Command of the Waters,* 37–60.

36. Ibid., 47–65.

37. Ibid., 60.

38. Ibid., 116.

39. National Water Commission, *Water Policies for the Future: Final Report to the President and to the Congress of the United States* (1973), 474–75, quoted in *Tribal Water Management Handbook,* American Indian Resources Institute (Oakland, Calif., 1988), 46.

40. For more on the conceptual development of nature and natural resources in European and Euro-American thought, see Clarence Glacken, *Traces on the Rhodian Shore* (Berkeley: University of California Press, 1973); and Neil Evernden, *The Social Creation of Nature* (Baltimore, Md.: Johns Hopkins University Press, 1992).

41. Michael L. Lawson, *Dammed Indians: The Pick-Sloan Plan and the Missouri River Sioux, 1944–1980* (Norman: University of Oklahoma Press, 1982), xxvi, 27–30.

42. Joseph H. Cash and Gerald W. Wolf, *The Three Affiliated Tribes (Mandan, Arikara, and Hidatsa)* (Phoenix, Ariz.: Indian Tribal Series, 1974), 2.

43. Roy W. Meyer, "Fort Berthold and the Garrison Dam," *North Dakota History* 35 (1968): 347–48. See also Arthur E. Morgan, *Dams and Other Disasters: A Century of the Army Corps of Engineers in Civil Works* (Boston: Porter Sargent Publisher, 1971), 41–42.

44. Cash and Wolf, *Three Affiliated Tribes,* 82–86; Morgan, *Dams and Other Disasters,* 41–57; and Lawson, *Dammed Indians,* 59–62.

45. Hearings before the House Committee on Interior and Insular Affairs, H.R. 2414, "Implementation of Certain Recommendations of the Garrison Unit Joint Tribal Advisory Committee," 102d Cong., 1991; Senate Report 102–250, "Implementation of Certain Recommendations of the Garrison Unit Joint Tribal Advisory Committee," 102d Cong., 1991. See also David L. Feldman, *Water Resource Management: In Search of an Environmental Ethic* (Baltimore, Md.: Johns Hopkins University Press, 1991), 88–89.

46. Lawson, *Dammed Indians,* 27–29, 55–58, 145–54.

47. Ibid., 29.

48. Hearings, House Committee on Interior; Senate Report 102–250; and Feldman, *Water Resource Management.*

49. Feldman, *Water Resource Management,* 179–90.

50. Alvis Little Eagle, "Tribes Want Plug Pulled on Corps Water Control," *Indian Country Today,* 14 October 1993.

51. Morgan, *Dams and Other Disasters,* 54.

52. Lawson, *Dammed Indians,* 46.

53. Ibid., 59–63, 91–92.

54. Quoted in Lawson, *Dammed Indians,* 179.

55. National Congress of American Indians, *Resolution,* no. 5, "Regarding Painted Rock Reservoir, Arizona," Annual Meetings, 14–18 February 1960, Denver, Colo.; Morgan, *Dams and Other Disasters,* 57–60; Lawson, *Dammed Indians,* 199; and Roger Moody, *The Indigenous Voice: Visions and Reality* (Utrecht, Netherlands: International Books, 1993), 187–209.

56. Meyer, "Fort Berthold," 348; and see also McCool, *Command of the Waters,* 177.

C. Richard Bath, Janet M. Tanski,

and Roberto E. Villarreal

THE FAILURE TO PROVIDE BASIC SERVICES TO THE *COLONIAS* OF EL PASO COUNTY

A Case of Environmental Racism?

In the mid-1980s, the national media in the United States focused on the problem of the *colonias* along the U.S.-Mexico border.[1] Attention was finally devoted to the hundreds of thousands of residents of these unincorporated rural developments with no zoning requirements, usually no water or sewage, no transportation facilities, or for that matter, little in the way of other basic services. The national media asked the question, How can such Third World conditions exist here in the United States? Indeed, although the *colonias* had existed for some time, many—including politicians in Washington, D.C., Austin, Texas, and elsewhere—seemed genuinely shocked to find such dismal conditions within the borders of the United States.

How did these *colonias* develop along the border, especially in the state of Texas? Since the overwhelming majority of residents were Hispanics or Latinos (in fact, the vast majority were either recent immigrants from Mexico or Mexican Americans), does the development of the *colonias* represent a case of racism in state and federal policy?[2] Specifically, does the failure to provide even minimum public services, such as water and sewage, represent a case of environmental racism? Were the Latinos in the *colonias* the victims of policies designed to keep them from obtaining the types of services most Americans regard as essential, simply because they were Mexican Americans? This chapter looks at the specific case of the *colonias* in Texas's El Paso County and whether the failure to provide basic water and sewage facilities constitutes a case of environmental racism.

The Political Background

El Paso became part of Texas by a vote taken in the early 1850s. Although it would have been more logical for the city to become part of New Mexico, as it was physically located within the boundaries of Nuevo Mexico under the administrations of both colonial Spain and independent Mexico, the vote was largely decided on the basis of support for slavery in Texas.[3] With secession, the defeat in the Civil War, and the reconstruction period, the politics of Texas revolved around race and the effort to deny Blacks equal protection under the law. As segregation became more and more a part of the economy, society, and politics of the state, Mexican Americans soon found themselves classified the same as Blacks and were denied the same rights and protections under the law.[4] By the 1950s, Mexican Americans, especially in the borderlands, were rigidly segregated and could look forward to a life with little educational or economic opportunity. Hotels throughout the state, for instance, carried signs that said, "No dogs or Mexicans allowed."

El Paso was not much different than the rest of Texas.[5] Although Mexican Americans were prominent in the early life of the community, much of the wealth of the Mexican American community had been taken by one means or another after both the Mexican and Civil Wars. In 1877, the famous Salt War—when Mexicans of El Paso briefly occupied the city—led to an Anglo attack on the political role of Mexican Americans, who, up until that time, had occupied some political posts and participated in elections.[6] After the Salt War, not one Mexican American was elected to a prominent city or county position until the 1950s. Mexican American voters were controlled and manipulated by various political rings, as well as by the constant fear of retaliation for voting by angry Anglos.

From the 1880s on, El Paso was a completely segregated city. Mexican Americans were ghettoized by housing, employment, and education. Both formal and informal practices by city officials led to this ghettoization. They were only permitted to buy lots or homes in certain parts of the city, they were harshly restricted in employment to those menial jobs at the bottom of the labor market, and they normally did not go beyond the sixth grade, which was regarded as the most education a Mexican American would ever require to perform his or her job well. During the 1920s, the city suffered from the onslaught of the Ku Klux Klan, which attacked the Catholicism of the Mexican Americans, took over the local school system, and renamed city schools for the Anglo heroes of the Alamo.[7] During the Depression, the federal government sent back to Mexico thousands of Mexican residents because of their alleged impact on the economy.[8] Many of those sent back were citizens of the United States, including residents of El Paso. Another disquieting event for Mexican Americans was "Operation Wetback" in 1954, when yet again, thousands were deported after a temporary halt to the *bracero* program, which the U.S. government set up to "contract" with the Mexican government for Mexican agricultural workers.

After World War II, however, many Mexican American veterans returned and refused to accept their second-class citizenship. Throughout the U.S. Southwest, groups such as the League of United Latin American Citizens were formed to improve the status and legal rights of Mexican Americans. In El Paso, the first Latino member of the city council was elected in 1951 and the first mayor in 1958. More rapid change did not occur until the 1970s, with the passage of the Civil Rights Act of 1964 and, especially, the Voting Rights Act of 1965, which allowed Texas minorities to challenge a host of institutional barriers to political participation. In the past, these barriers had included the poll tax, rigorous voting registration requirements, and gerrymandered and at-large districts. Gradually, under the Voting Rights Act, many of these institutional barriers were removed by the Department of Justice or through court action, and by the 1980s, Mexican Americans were voting with little difficulty and increasingly electing their own representatives.[9]

The strong political and economic elite that had dominated Texas and El Paso politics since the 1930s was severely challenged in the 1980s by legal mandates, the changing nature of the state's population and economy, and the political involvement of women and minorities. By the end of the 1980s, the power of that elite was broken in both El Paso and Texas, and a new era of politics began with the rise of the Republican Party in a now genuinely two-party state and the increasing participation of groups that had previously been disenfranchised or not represented in the political system.[10]

Moreover, a series of legal challenges to institutional racism began to change the overall structure of Texas during the 1980s. A Mexican

American Legal Defense Fund suit against the University of Texas system showed that the border universities, including the branch in El Paso, were systematically underfunded, and as a result, the court ordered a redress of grievances. Similarly, court action required a change in public school financing, which had favored the richer school districts, and in the process, schools in the El Paso region were found to be among the most disadvantaged in the state.[11] In 1994, a court of inquiry under a state judge looked at El Paso in comparison with the rest of Texas in such areas as highway construction, mental health, and child care and found that El Paso always received far less state funding than it was entitled to based on the size of its population.[12] These imbalances, which the 1995 state legislature promised to correct, represent a tradition of discrimination in public policy against Mexican Americans in Texas. To this day, little restitution has been forthcoming. The question is: Does the failure to provide water and sewage to *colonia* residents also embody this tradition of discrimination?

Background to the Housing Problem

There are several reasons why *colonias* developed in El Paso County. First, about two decades ago, a rapid population increase as a result of stepped-up immigration from Mexico[13] created an enormous demand for housing. El Paso has long been a doorway to the United States for Mexico's immigrants. For over a century, recent immigrants lived in the slums of Chihuahuita or South El Paso before they moved on to other areas of the city or the rest of the country. The number of immigrants has substantially increased, however, since the 1970s, and the city has had great difficulty finding housing for the newcomers. By the early 1970s, some 18,000 residents of South El Paso lived in abysmal conditions.[14] In response, the city began a Tenement Eradication Program, which put even greater pressure on existing housing.[15] Yet because of the economic collapse of Mexico after 1982, there were even larger waves of immigrants coming to the city in the 1980s. By 1990, 25 percent of El Paso's residents were foreign born, almost all of whom came from Mexico; in the *colonias*' region, 96 percent of the inhabitants came from Mexico.

A second major factor is the poverty found in the border region.

El Paso, along with other Texas border cities such as Brownsville, McAllen, and Laredo, always ranks in the bottom five of all Metropolitan Statistical Areas in the United States on poverty scales. Per capita income in El Paso County in 1992 was $12,307, or only 62 percent of the national figure of $19,802. In South El Paso, the per capita income was actually below $10,000, making it the poorest census tract in the city.[16] Unemployment in the region is generally double that for the rest of the nation (although for the first time in many years, the advent of the North American Free Trade Agreement has considerably improved the employment rate for the city). As a result, 155,298 people or 26 percent of all residents in 1990 were classified as poor by federal government standards, and of that amount, 88 percent were Hispanic.[17]

Poverty, in turn, made it nearly impossible for many in El Paso to purchase their own home. It was estimated that 70 percent of all persons desiring to buy homes in the city in 1976 could not do so because they could not qualify for home loans. By 1988, however, the number who tried to qualify for home loans and could not receive them was reduced to 39.4 percent.[18] As the tenements in South El Paso were eradicated, developers in the rural areas began to recognize that they could sell lots to the unsuspecting and promise them all sorts of services that would never be delivered. For the poor, the chance to own land and build homes was a dream come true. They rarely challenged the terms or conditions under which they bought the undeveloped land, and, it should be added, they rarely ever actually obtained title to the land.

As recently discovered, there was another side of this picture. A 1992 study by the Hispanic Leadership Institute of El Paso of housing loans disclosed that turndown rates for Hispanics, Blacks, and whites were respectively 36.9, 34.8, and 19.3 percent. When applicants above the median family income were tested, it was found that the turndown rate was 31 percent for Hispanics, 25.9 percent for Blacks, and 14.1 percent for whites.[19] In response to this study, major efforts were made to correct discrimination in home loans and, by late 1994, it was reported that although the turndown rates were still high, it was easier for minorities to obtain home loans in El Paso.[20] In the $50,000 to $60,000 income range, for instance, the rates of loan approvals were respectively 79, 82, and 81 percent for Hispanics, Blacks, and whites. In the $20,000 to $30,000 income range, however, it was 79, 82, and 81 percent respectively. Needless to say, even these improvements came far too late to help the thousands who, during the 1970s and 1980s, could only purchase land and begin the process of building their own homes in the rural *colonias.*

The Politics of Water

Historically in El Paso County, there has always been a contest between those who wanted to rely on surface water provided by the Rio Grande River and those who wished to rely essentially on groundwater provided by two aquifers, the Hueco and Mesilla Bolsons. When the Rio Grande project began in the early part of this century and the river was dammed at Elephant Butte, the city of El Paso opted not to participate and decided instead to rely on groundwater for future consumption. Area farmers formed the El Paso County Water Irrigation District No. 1 to provide river water from the Rio Grande project to farms in the region. Over the years, there has been a constant struggle between the district and the city over water rights and the perceived threat to the farmers of encroaching urbanization.

Up until the 1950s, the city's water board depended on both surface and groundwater sources, but in 1952, newly elected Mayor Fred Hervey formed the El Paso Water Utility (EPWU)—governed by the Public Service Board (PSB)—to operate the water and sewage facilities of the city "with the same freedom and in the same manner as are ordinarily enjoyed by the Board of Directors of a private corporation operating properties of a similar nature."[21] Since the PSB decided to rely almost exclusively on groundwater, by the late 1970s, El Paso received less than 10 percent of its water from the river.

From the start, the EPWU/PSB was regarded essentially as a tool for real estate developers to control the growth of the city. Water was to be utilized primarily to ensure continuous growth, and little attention was paid to the future supply of water or the need to conserve a scarce resource in a semiarid region. El Paso had the lowest water rates of any city in the entire Southwest; indeed, the more water used by the consumer, the lower the rate paid. By 1980, little attention had been paid to future water supply or distribution, since the farmers were well served by the district, real estate interests controlled water for the city, and the public was pleased with extremely low water rates.

The PSB was appointed by the city council, and the mayor was automatically a member of the board. Of a total of twenty-one members of the board up to 1989, fourteen were directly tied to real estate and development interests. During the most critical years for the EPWU/PSB, when decisions to provide water to outlying areas had to be made in the 1980s, four out of five members of the PSB were connected to real estate development. All three Mexican American members of the PSB prior to 1989 had close ties to development interests.[22] In other words, the agency chiefly responsible for providing water to the city and surrounding areas was basically operating as a "captured agency" for real estate interests.

The EPWU/PSB also operated out of the public eye. Much of the decision making was in the hands of the professional staff running the water utility, and even disgruntled board members bowed to them. As one former PSB member commented, "We were told where subdivisions were going to go and to bring water lines. There was no dialogue at all."[23] Neither the city council nor any mayor ever seriously challenged decisions made by the EPWU/PSB. Indeed, during much of the time from 1952–1990, elected city officials were constantly turning over (only two mayors were reelected), making it especially difficult for them to keep up with the decisions reached by the water board.

By 1980, the EPWU was regarded as one of the soundest utilities in the country; in 1987, it had a total indebtedness of $51,790,000, an income of $39,665,193, and assets totaling over $316 million. Quite clearly, the EPWU was making money. But in 1978 a decision was reached by the PSB not to extend water or sewage lines beyond the city limits. Ostensibly, this was because of the expense involved. The EPWU staff had bad memories of a previous annexation of Ysleta in the lower valley that had cost a great deal of money and the city was reluctant to extend extraterritorial jurisdiction to the poorer areas of the county.

The actual reason for not extending water and sewage lines (which, incidentally, were extended in a few cases to wealthier developments) was that El Paso had filed an application with the New Mexico State Engineer to drill 326 wells in the Mesilla Bolson to extract water from across the New Mexico state line. Naturally, New

Mexico resented this attempt, and a long and bitter struggle for water in the Mesilla Bolson finally culminated in a mutual agreement in 1991. El Paso agreed to surrender any claim to Mesilla groundwater in return for more surface water from the Rio Grande. As a result of this legal/political battle with New Mexico, water lines were not extended to the *colonias* in the county for most of the 1980s.

Water policy in Texas has traditionally been dominated by rural interests and, consequently, governmental structures reflect farmers' interests. Also, Texas has no effective groundwater law and County Commissioners Courts, which serve as an advisory board, have no policymaking powers, and cannot mandate the provision of services to outlying areas. Since the city would not extend its jurisdiction, and the county had none, the most logical candidate (and the one with the largest funds) to supply water and sewage services was the EPWU/PSB.[24] But there was no political pressure on the city utility to provide such services, and it did not want to voluntarily undertake such a huge financial commitment.

A political push was finally provided by the El Paso Interreligious Sponsoring Organization (EPISO), which was based on Saul Alinsky's Industrial Area Foundation's organization to help the poor. EPISO was founded in 1977 and closely tied to the Roman Catholic Church, enjoying the full support of El Paso's bishops. Its major goal was to organize the Mexican American community in El Paso through political participation.[25] EPISO's mass rallies, for instance, which attract almost mandatory attendance by those running for office, allow public issues to be critically discussed and solutions sought. Early in the 1980s, when EPISO began to stress the need for provision of water and sewage facilities to the *colonias,* it rapidly became a major political issue in El Paso and Texas politics. Governor Mark White led a contingent of Texas officials that eventually included such prominent public figures as Jim Hightower, Gary Mauro, and Ann Richards. The *colonias* attracted the attention of the national media, and politicians from Austin to Washington, D.C., soon called for an end to these miserable conditions.

In 1985, a Lower Valley District Water Authority was created to provide water to the *colonias* of that region, but it had neither funds nor water. The PSB finally agreed to build a water treatment plant for the city of Socorro in 1988, and since then, numerous efforts have been made to provide water and sewage services to *colonia* residents. Both the federal and state governments have supplied funds, and money is far easier to obtain than in the past.

Surprising for Texas, three of the most critical decision makers involved in the *colonias* issue were women. Suzie Azar, who opposed the water suit against New Mexico, ran successfully for El Paso's mayor's seat in the 1989 election, in which water was an issue. In 1991, she finally accepted the agreement that ended the New Mexico dispute and permitted the city to look for other means to secure its future water supply. Around the same time, a series of stories on the PSB by Peter Brock in the *El Paso Herald Post*, which pointed out the insularity of the board and its secret decision-making processes that favored real estate developments, forced several members to resign, since they had already gone well beyond their legal term of office. Mayor Azar replaced them with more publicly conscious representatives, including two Mexican Americans. Moreover, Alicia Chacon became the first Latino El Paso county judge in over a century, and the election of liberal Democrat Ann Richards as Texas governor in 1990 also interjected a more sympathetic ear to the plight of the *colonias* in Austin.

More important changes occurred at the EPWU in 1989, where a relatively young Mexican American, Edward Archuleta, replaced the former executive director who had been in office since the EPWU/PSB was established. Archuleta began a serious effort at conservation, something that had only been given lip service under the previous director. It was this genuine attempt to start stringent conservation programs that allowed for the eventual settlement of the dispute with New Mexico. New Mexico had previously argued—and believed—that El Paso was not interested in water conservation but simply wanted to deplete New Mexico's water sources.

In 1993, a bill was introduced in the Texas legislature to provide for one water planning agency for El Paso County. Under this bill, the EPWU/PSB would become a countywide planning agency, supplying water and sewage facilities to all county residents. The bill was defeated because of the political leverage wielded by the Texas Water Commission. A similar bill was introduced in early 1995, which passed with little opposition, but the end result was the creation of a politically weak EPWU with very little planning power. Still, it is foreseeable that in the near future the *colonias* of El Paso will finally enjoy the same kinds of services that almost all Americans do.

Conclusion

The failure to provide essential services to the residents of the borderlands *colonias* became a national issue during the 1980s. In the case of El Paso County's *colonias,* the question raised was whether the failure to provide services, specifically water and sewage, was the result of environmental racism. Given the past segregation and discrimination of Mexican Americans found in Texas and El Paso, there is sufficient historical evidence to support the proposition that the development of the *colonias* was the result of past institutional racism. Further, Mexican Americans had little or no opportunity to participate in the political system. They not only had no representation, they were not even present in the pre-1960 Texas political system.

The growth of Mexican American interest groups, the Civil Rights Act of 1964, and Voting Rights Act of 1965 all provided means for political representation for the Latinos in the borderlands region. In turn, increasing political representation and more strident political voices have ensured that Mexican Americans in Texas will no longer be left outside the political arena. Political representation has also brought the recognition that Mexican Americans should be participants in the American dream of upward social, economic, and political mobility.

In the case of El Paso, continued immigration and poverty created housing shortages that were answered by the growth of unregulated (and unscrupulous) developments. Virtually the *only* housing available to newcomers was in the *colonias,* where they could own their own homes. Were private and public policies restricting housing availability the result of racism? Probably not, although the failure to provide home loans to similar income-level minorities is certainly a disquieting factor. More than likely, housing development simply followed economic realities, where developers make far more money on upper-income homes than on houses for the poor.

Was the failure to provide water and sewage facilities the direct result of racist policies? Again, probably not, since some harsh economic realities had to be faced. How would these poorer residents possibly pay for the provision of such services? They cannot, unless there is substantial support from all levels of government, and the PSB had learned from past experience that they could not expect to make a profit—as they were required to do—providing such services to outlying areas. The costs were prohibitive given the distances involved between *colonias* and individual homes.

The victims of the failure to provide water and sewage facilities were poor Mexican Americans, many of whom migrated to the United States for an opportunity to own a home. If the immigrants were Asians, they would have encountered the same housing problem, but they might not have encountered the history of institutional racism that has severely curtailed the ability of Latinos in Texas to ascend the social and economic ladder. One thing is certain: although there is some evidence of economic and political advancements, as long as immigration continues unabated and those immigrants are largely Hispanic, they will face the economic hardships encountered by previous immigrants, and the situation will not be much different.

Notes

1. Texas Department of Human Services, *The Colonia Factbook: A Survey of Living Conditions in Rural Areas of South and West Texas Border Counties* (Austin: Texas Department of Human Services, 1988); and U.S. General Accounting Office, *Rural Development: Problems and Progress of Colonia Subdivisions Near the Mexican Border* (Washington, D.C.: General Accounting Office, November 1990).

2. Here, the definition of racism is that employed by the United Church of Christ Commission for Racial Justice, *Toxic Wastes and Race in the United States: A National Report on the Racial and Socioeconomic Characteristics of Communities with Hazardous Waste Sites* (New York: United Church of Christ, 1987), x.

3. Wilbert H. Timmons, *El Paso: A Borderlands History* (El Paso: Texas Western Press, 1990), 116–33.

4. David Montejano, *Anglos and Mexicans in the Making of Texas, 1836–1986* (Austin: University of Texas Press, 1987).

5. Mario T. Garcia, *Desert Immigrants: The Mexicans of El Paso, 1880–1920* (New Haven, Conn.: Yale University Press, 1981); and Oscar J. Martinez, *The Chicanos of El Paso* (El Paso: Texas Western Press, 1980).

6. Leon Metz, *Border: The U.S.-Mexico Line* (El Paso: Mangan Books, 1989), 319; and Timmons, *El Paso*, 165–66.

7. Shawn Lay, *War, Revolution, and the Ku Klux Klan: A Study of Intolerance in a Border City* (El Paso: Texas Western Press, 1985).

8. Oscar J. Martinez, *Border Boomtown: Ciudad Juarez since 1848* (El Paso: Texas Western Press, 1978), 80.

9. Howard D. Neighbor, "Latino Political Participation in a Bicultural Setting," in *Latinos and Political Coalitions: Political Empowerment for the 1990s*, ed. Roberto E. Villarreal and Norma G. Hernandez (New York: Greenwood Press, 1991), 9–18.

10. C. Richard Bath, Janet M. Tanski, and Roberto E. Villarreal, "The Politics of Water Allocation in El Paso Colonias," *Journal of Borderlands Studies* 9, no. 1 (spring 1994): 15–38.

11. Robert E. Webking and Gregory R. Rocha, *Education in Texas* (Minneapolis, Minn.: West, 1994).

12. Thomas Black, "State Officials: City Road Planning Poor," *El Paso Times*, 16 November 1994, 7A; and Gary Scharrer, "Road Chiefs: El Paso Needs Less Money for Repairs," *El Paso Times*, 17 November 1994, 1A.

13. Bath, Tanski, and Villarreal, "Politics of Water Allocation," 17–18.

14. James W. Lamare, *An Evaluation of the Tenement Eradication Program of the City of El Paso* (El Paso, Tex.: Department of Planning and Research, December 1974).

15. Benjamin Marquez, *Power and Politics in a Chicano Barrio* (New York: University Press of America, 1985).

16. U.S. Department of Commerce, Bureau of Economic Analysis, *Machine Readable Data Files* (Washington, D.C.: Government Printing Office, 1992).

17. Bureau of the Census, *1990 Census of Population and Housing, Summary Tape File 3C* (Washington, D.C.: Government Printing Office, 1993).

18. City of El Paso, *Comprehensive Housing Affordability Strategy for El Paso, Texas* (El Paso, Tex.: Department of Planning, Research, and Development, October 1988).

19. Hispanic Leadership Institute of El Paso, *Mortgage Rates: Acceptance and Turndowns* (El Paso, Tex.: Hispanic Leadership Institute of El Paso, 1992).

20. David Crowder, "Minorities Given More Home Loans," *El Paso Times*, 28 December 1994, 1A.

21. Conrey Bryson, "El Paso Water Supply: Problems and Solutions" (Master's thesis, Department of History, Texas Western College, 1959), 27.

22. Peter Brock, *El Paso Herald Post,* 18–22 April 1988, series of articles exploring the nature of the PSB.

23. Brock, *El Paso Herald Post,* 20 April 1988, A9.

24. Parker, Smith, and Cooper, Inc., *Water and Wastewater Management Plan for El Paso County, Texas,* report prepared for the El Paso City/County Health Unit, May 1988.

25. Roberto E. Villarreal, "EPISO and Political Empowerment: Organizational Politics in a Border City," *Journal of Borderlands Studies* 4, no. 2 (fall 1989): 75–89.

III

Confronting Environmental Injustices

John G. Bretting and Diane-Michele Prindeville

ENVIRONMENTAL JUSTICE AND THE ROLE OF INDIGENOUS WOMEN ORGANIZING THEIR COMMUNITIES

In recent years, a body of literature has developed around the concepts of *environmental injustice* and the emergence of a burgeoning social justice movement. Other chapters in this volume have given considerable attention to these topics. During the 1980s, community-based grassroots environmental organizations formed in opposition to the problems of pollution caused by local businesses.[1] These new organizations, which are smaller in size than national ones, tend to represent lower-status and lower-educated individuals.[2] It is estimated that between 6,000 and 10,000 local groups now exist in the United States.[3] This grassroots mobilization has grown in response to local environmental threats and to the marginalization of working-class, indigenous community interests by the largely white, middle-class environmental movement.[4]

Studies of the political behavior of women and working-class people in indigenous communities provide a rich source of information and reinforce the legitimacy of grassroots activism as a valid form of political expression.[5] In this chapter, we present findings from our study of indigenous Chicana and Native American women who have mobilized their communities to promote environmental justice in New Mexico.

Environmental Justice

Environmental problems in New Mexico are like environmental problems elsewhere. Those individuals and groups with greater power influence the uneven distribution of hazards. In other words, there tends to be a relationship between social class, race, and gender and the exposure of individuals to environmental problems.[6]

When we refer to environmental injustice we mean situations where a community's citizenry perceives that the local, state, or tribal government is failing to protect their lives and property from environmental pollution and its associated costs.[7] "Environmental justice refers to the belief that both environmental benefits and . . . costs should be equally distributed in society, and that corporations should be obligated to obey existing laws, just as individuals are so obligated."[8]

Indigenous Communities and Women Organizers in New Mexico

Although the leadership role of women in the development of social movements has been documented, the literature has yet to focus on the particular role of Chicanas and Native American women as community organizers. Using qualitative data from personal interviews, we examine the motives, goals, and strategies of twelve women of color who are leaders in environmental justice organizations in New Mexico; discuss their commonalities and differences, as well as the approaches to organizing used by these indigenous women; and compare their strategies to those of "traditional" environmental organizations (see Longo, next chapter, for an extended discussion of the relationship between nontraditional and traditional environmental organizations fighting environmental injustice). The research questions addressed include:

- What is your organization's purpose or mission?
- How does your organization do its business?
- What are the various strategies your organization uses?
- How are decisions made within your organization?
- Have you had an impact within your community?
- Are you a feminist? Environmentalist? Political activist? Community organizer?

The findings suggest that these indigenous activists are (1) devoted political actors, who (2) work through structurally and ideologically democratic organizations, to (3) improve the social, environmental, and economic conditions of their constituents, and to (4) influence public policies affecting their communities. Their specific concerns include those that arise from environmental conditions and community development, neighborhood safety, availability of low-income housing and public transportation, employment opportunities, and workplace hazards.[9] According to Robert Bullard's studies of grassroots environmental activism in Black communities, issues most likely to attract citizen support include those that (1) endorse policies that favor the disenfranchised; (2) concentrate on equality and distributional impacts; (3) advocate direct action and are ideologically democratic; (4) consider the economic needs of the community; and (5) solicit the support of local civic and religious groups.[10] Toxins, pesticides, water quality and groundwater contamination, soil erosion, waste reduction, and incineration are also specific environmental concerns of indigenous and working-class people.[11] Each issue relates directly to the quality of life in both economically depressed urban areas and rural communities.[12]

Race, Gender, and the Social Science Literature

The compartmentalization by political scientists of the study of community-based political (and social) movements results in the marginalization of this area of research into the subfields of urban politics, race and ethnic politics, and women and politics. For example, the social science literature has traditionally treated women's collective action as an extension of their domestic responsibilities rather than as legitimate political activity motivated by class, race, economic, or other concerns.[13] Feminist scholar Sandra Morgen argues that

> women's community-based political activism is a conscious and collective way of expressing and acting on their interests as *women,* as *wives and mothers,* as *members of neighborhoods and communities,* and as *members of particular race, ethnic, and class groups.* To collapse a complex political consciousness into the more narrow confines of domestic values, interests, and roles is to distort both the motivation and the political implications of this mode of resistance.[14]

The social science literature has adequately discussed the connection between women and race. Several scholars have codified the term *twice a minority* to reflect the double stigma that Native American women and Latinas face socially, economically, and politically. Yet the intersection between gender, race, and the environment has received limited and indirect attention in the ecofeminist literature. As Karen Warren notes in her essay on feminism and ecology, a large portion of this literature lacks a racial dimension.[15]

Our research, discussed below, constructs a linkage and illustrates the relationship for Native American women and Latina environmental activists between environment, race, and gender. These women, in most instances, are engaged in macroscopic struggles to remedy social concerns, including establishing a healthier and safer environment for themselves and their communities. Their identity as indigenous women, ethnic and racial minorities, and environmental activists is predicated on their personal ethics and shared belief systems. These, of course, transcend any oversimplification and labeling, such as "feminist" and/or "environmentalist." As these women explain, it is difficult to pigeonhole one's lifework.

Learning from the Field: Interviews with Activists

RESEARCH DESIGN

Twelve self-identified indigenous women of color (Chicanas and Native American) who are leaders in different community-based environmental organizations in New Mexico were interviewed. A snowball sample, based on reputation in the environmental organizing community, was employed to identify the study participants. The small sample (N = 12) prevents any generalizations to the larger population of environmental justice activists, but it does allow for detailed descriptions of the goals and strategies used by these women to promote environmental justice in New Mexico. Although comparisons across jurisdictions cannot be made, the research approach presented may serve as a model for others interested in studying grassroots organizations. The research findings can also assist other local groups in the formative stages.

The interview guide (see appendix A) was designed to: (1) generate a history of the activists' participation in the environmental justice organization, (2) describe their role and influence in decision making, and (3) examine their strategies and tactics. The activists were interviewed in person, either at their organizations' office, their home, or other public places in the greater Albuquerque, New Mexico, metropolitan area. The interviews took between forty-five and ninety minutes; three of them were conducted as a pilot study during November and December 1991. The remaining nine were conducted between October and December 1994. To guarantee the confidentiality of the respondents and their organizations, pseudonyms were used.[16]

QUALITATIVE FIELD RESEARCH APPROACH

Qualitative methods are often more useful than quantitative ones for investigating the hows and whys of human action.[17] This is especially relevant when the researcher has a limited population from which to draw a sample—as in the case of political elites. Long, open-ended interviews give the respondent nonstandardized, special treatment. They stress the participant's definition of the problem, and allow her to communicate her notion of relevance.[18] More strikingly, qualitative research methods look for specific patterns of interrelationship between many categories. The patterns discovered assist in the development of theory and provide an understanding of the phenomena under investigation.[19] The research goal here is to develop as rich and accurate a profile as possible for each of the indigenous activists. Data are collected, focused, and analyzed.[20]

The research findings are reported in two sections. Section one, "Indigenous Women Organizers Promoting Environmental Justice," profiles the activists and their community-organizing experience. Section two, "Political Activism from the Grassroots: Mobilizing Communities for Change," examines their specific organizational goals and decision-making processes, and provides a detailed analysis of organizational strategies.

Indigenous Women Organizers Promoting Environmental Justice

DEMOGRAPHIC PROFILES OF COMMUNITY LEADERS

According to activist Lois Gibbs, grassroots leaders are generally working-class people with twelve years or less of school, who lack "any formal organizing training, have not been involved in any

Table 1. Demographics

	Race	Education (Years)	Total Income	Age	Children at Home
Carmen	Indian	13	$35,000	53	0
Dalia	Chicana	14	$70,000	43	1
Donna	Indian	11	$5,000	56	0
Felicia	Chicana	17	$45,000	30	0
Jackie	Indian	16	$36,000	24	0
Juana	Chicana	16	—	48	2
Leslie	Indian	12	$34,000	31	0
Linda	Hawaiian	19	$45,000	46	0
Monica	Chicana	10	$21,000	60	0
Rose	Indian	21	$16,000*	37	3
Susana	Chicana	10	$35,000	50	2
Terri	Chicana	12	$7,000*	47	0

*In 1991 dollars

other social justice issue and come from and live in a very 'traditional' kind of lifestyle."[21] Only a minority of the leaders interviewed here, however, fit that profile (see table 1). Instead, they fall into three categories: "lifelong activists," "young professionals," and "traditional organizers."

The five "lifelong activists" have been involved in grassroots organizing for twenty years or more (see table 2). They had mentors and were politicized in their teens or early twenties. They were influenced, substantially, by the civil rights and women's movements and by indigenous peoples' uprisings. Their political ideology was formed at that time and reflects a communitarian worldview of social justice and environmental stewardship. Four of these women hold paid positions in their organizations, the fifth owns a business through which she donates work and resources to two indigenous environmental groups. Four of the five have some college; in total, they average sixteen years of education. All but one of the leaders are mothers and all but one have live-in partners. Finally, all of the "lifelong activists" see their work as a calling rather than an occupation or avocation.

Three of the leaders interviewed are classified as "young professionals." They are in their twenties or early thirties. Each deliberately sought work in a grassroots environmental organization to serve her indigenous community. In general, these women earn

Table 2. Ideology

	Feminist	*Environmentalist*	*Political Activist*	*Community Organizer*	*Years Involved*
Carmen	No / X	Yes / O	Yes	No	20.0
Dalia	X	Yes	Yes	Yes	5.0
Donna	No	O	Yes	Not yet	2.0
Felicia	Yes	O	Yes	Yes	4.5
Jackie	X / O	Yes	No	Not yet	8.0
Juana	No / O	No / O	Yes	Yes	48.0
Leslie	No / O	Yes / O	O	Not yet	1.5
Linda	Yes	Yes	Yes	No	25.0
Monica	No / X	Yes	Yes	Yes	5.0
Rose	No / X	No / O	Yes	Yes	20.0
Susana	Yes	Yes	Yes	No	2.0
Terri	No / O	No / O	No	Yes	21.0

X: These women reject the term *feminist;* however, they believe in equal rights and that individuals are responsible for their own success.
O: These women object to Western definitions of *environmentalist, feminist,* or *political activist.* Instead, they propose communitarian approaches that reflect the culture and belief systems of indigenous peoples.

higher salaries than their cohorts and hold technical positions. Two of the three have college degrees in environmental science. Two have live-in partners; none have children. On the surface, these women resemble the paid professionals of the traditional environmental groups. Their primary identity, however, is that of indigenous women struggling for justice for their communities.

The four remaining women are "traditional organizers" who see themselves less as leaders than as community problem solvers. They have been active in community politics for two to five years. Each woman became active politically because of a specific incident that dramatically affected her health and/or her community's environmental well-being. Of the four women, one is employed by her organization, one works in the public sector, and two are retired because of disabling illnesses acquired from workplace toxins. One has two years of college; the other three did not finish high school. All are mothers; two are grandmothers. Three have dependents living with them. The "traditional organizers" fit the profile of individuals reacting to a specific problem or issue in their community.

Five of the women interviewed identified themselves as members of North American Indian nations, six are Chicanas (of Mexican

American heritage), and one is a Hawaiian native. Eleven of the twelve referred to her people as indigenous or Third World. The youngest leader was twenty-four when interviewed; the oldest was sixty. The mean age was forty-four. Eight of the women hold paid staff positions in their organizations. Although their average total household income fell in the low thirties, the leaders' annual incomes ranged from approximately $2,000 to $36,000. Most of those doing paid work earn salaries up to the low twenties. None are motivated by financial gain.

Community Mobilization: The "Indigenous Feminist Approach"

All of the leaders interviewed projected passion and commitment to their work, and acknowledged their influence in organizational decision making. They described their role in the community as "problem solver," "organizer," "activist," or "advocate" (see table 2). Each sees her community work in terms of her life purpose, inseparable from her identity. All support equal rights for women and many spoke with pride of empowering others. They expressed concern for their working-class sisters who manage multiple roles and responsibilities in their daily lives. Several respondents recognized that they are role models. Despite their avowed support for the social and political advancement of women, however, only three of the twelve leaders identified herself as a feminist. Like many Third World women, they distrust and reject white, middle-class feminism.[22]

The nine women who rejected the label "feminist" generally fell into one of two groups. The "individualists" believe in equal rights for all, and they see themselves as strong, independent, and capable women responsible for their own success. The "indigenous feminists" object to what they call Western definitions of feminism. Instead, they propose a communitarian approach that reflects the culture and belief systems of indigenous peoples. This holistic approach is founded on consensus building; it incorporates the family and, by extension, the community.

Similarly, five of the twelve respondents rejected the label "environmentalist," claiming that the mainstream environmental movement is out of touch with the economic and social realities of the poor, the working class, and people of color. They prefer to identify themselves as Third World or "indigenous environmentalists in the environmental justice movement." The principal differences between these movements are their approach, constituencies, and philosophy. The environmental justice movement is concerned with the economic and political empowerment of Third World communities (See White, this volume, table 3).

Encompassing a broad scope, this grassroots movement incorporates social concerns so that exposure to chemical hazards in the workplace, for example, is not simply a labor issue but a class and racial issue. All of the women interviewed believe that Third World communities are specifically targeted for placement of polluting industries and facilities. They see the toxic contamination of their communities as systematic genocide. Their involvement in environmental issues is highly personal because threats to the environment are interpreted as threats to their families and communities. The environmental ethic held by these women is integral to their spirituality and inseparable from themselves. As one woman explained:

> Women are life-givers, their connection to Mother Earth is important. Women feel the impact of environmental issues more because they are in the home. Women look after the health and safety of the family, and by extension, the community.[23]

Political Activism from the Grassroots: Mobilizing Communities for Change

Each of the grassroots organizations represented by the women interviewed is structurally egalitarian and democratic. Some strive for political and economic transformation of the community by increasing public awareness and access to information (see appendix B). All seek to empower community members through leadership training and/or education. Each organization actively promotes consensus building, shared decision making, and community participation in goal setting and policy formulation. In the process, community members exercise self-determination and learn advocacy skills. As other researchers have found: "Grassroots organizations growing up around local toxic pollution . . . issues have different motivations and different sorts of leadership than national organizations."[24]

Several groups actively recruit their membership to promote gender equality and ethnic/racial diversity, and to ensure representation of poor and low-income citizens in their leadership. Ethnic identity and issues of race are highly salient for eleven of the twelve women interviewed. Working successfully across cultures is a shared organizational goal. The majority of these groups work at improving racial and ethnic relations by celebrating community heritage, participating in cultural exchange programs, and developing cooperative relationships with other groups. Some of the environmental organizations represent specific indigenous communities. All of the groups maintain a large number of women in leadership positions.

Organizational Goals and Decision Making

In general, the grassroots organizations represented here seek to clean up contaminated sites or otherwise improve local environmental conditions (see table 3). Their concerns range from the preservation of cultural and religious artifacts to workplace safety. Although their scopes extend from solving neighborhood-level problems to forming networks with organizations internationally, their common focus is the environmental health and safety of individuals within the community. These groups share similar organizational characteristics, but go about their business in different ways. For example, some focus their efforts on negotiating with industry and government agencies for pollution abatement and site cleanup; others on program development for the prevention of soil and groundwater contamination; still others use community land use planning or coalition building to achieve their goals.

The goals of these grassroots organizations can roughly be divided into five categories (see appendix B, "Goals Legend"). Three of the groups that specifically serve indigenous communities work to provide technical assistance and information to their constituencies to assist them in developing environmental programs. Three of the organizations with a local scope advocate community environmental health and seek to provide reparation to victims of toxic poisoning. Two of the groups with a statewide or regional scope function as think tanks to increase citizen participation in environmental policymaking. The two organizations that employ the greatest number of different strategies work to empower citizens to achieve racial and gender equality, as well as social, environmental, and economic justice. The two groups with an international scope cross national borders to advocate land, water, environmental, religious, and cultural rights for indigenous communities in North and Central America.

The organizations represented here either have paid staffs headed by an executive director (or codirectors) who answers to a board of directors, or they have corporate-like structures with elected officers who volunteer and answer to the general membership. All are private, nonprofit organizations with the exception of one group that operates as a citizens committee authorized by a tribal village council. Each of the twelve groups formulates policy and reaches decisions consensually.

Organizational Strategies and Tactics

The grassroots groups employ various strategies to influence policy. Their activities include registering voters, organizing conferences around environmental justice issues, publishing technical reports, and securing neighborhood improvements. The mature organizations have greater technical and financial resources, established working relationships with government and industry, a track record of successes, and employ many more strategies than fledgling groups. The newest organization studied here is two months old; the group operating the longest was formed in the mid-1970s. In total, the organizations use over twenty different sets of strategies and a multiplicity of tactics to achieve their goals (see appendix B, "Strategies Legend").

The strategies employed by the grassroots groups generally fall under one of five headings. "Strategies to Educate and Inform" incorporate community leadership training, holding public seminars and workshops, developing youth programs, producing publications, registering voters, and hosting candidate debates. "Community Relations Strategies" include winning celebrity endorsements, producing audiovisual aids, holding public hearings or demonstrations, exploiting media opportunities, and organizing boycotts. "Partnership Strategies" involve forming alliances or coalitions with civic or religious groups, labor, industry, or government, as well as networking to share resources with other grassroot groups. "Strategies to Develop Resources" include fund-raising, developing credible research reports, acquiring and/or providing expertise, and employing experts as advocates or spokespersons. Lastly, "Strategies Using the System" incorporate monitoring regulatory enforcement and compliance, lobbying political or bureaucratic officials, and litigating to influence public policy.

Influencing Public Policy: Community Impacts and Outcomes

Students of public policy have long sought to understand how and why groups organize to affect social and/or political change, the processes by which public policy is formulated, and how decisions are made.[25] Group theory views politics as a mosaic of interacting groups. Problems are identified in the context of group life. As groups seek solutions to their problems, they "lobby" government to adopt their preferred position. The favored approach within the discipline, the pluralist theory, characterizes the U.S. political system as one in which all active and "legitimate" groups can influence the public decision-making process. Some groups, however, are inevitably excluded. Hence, the emergence of environmental justice activists outside the formal institutional political arena to represent marginalized groups' interests (see Camacho, at the beginning of this volume, for a more extensive discussion).

The organizations examined here enjoy varying degrees of success (see appendix B, "Organizational Attributes" and "Impact Legend"). All have managed to solicit the participation of community members in identifying and formulating solutions to common problems. Ten groups have influenced environmental policy either formally, by lobbying for passage of legislation or enforcement of regulations, or informally, by applying pressure on government or industry to address community needs. Half of these organizations have won litigation or legal settlements that impact substantially on community resources or individuals' rights. Five have succeeded in preserving or repatriating indigenous peoples' land, resources, or cultural or religious artifacts. All the organizations have acquired funding and/or official authorization to develop or operate programs. Eight groups have implemented environmental cleanup projects.

The youngest groups, of course, have had the least time to achieve their goals. The mature groups that have made an impact see their mission as part of a process. Their general purpose, achieving social justice and economic equity for their constituencies, is unlikely to be realized any time soon. Although battles are won, new problems emerge. Hence, the grassroots organizations continue to play an important role in the community, organizing and unifying citizens, representing their concerns, and striving to meet community objectives. To a large extent, these groups fill a need that is not being met by elected politicians.

Political Activism: An Indigenous Perspective

Our findings are similar to a study conducted by Nicholas Freudenberg and Carol Steinsapir,[26] which indicated that leaders of grassroots environmental justice organizations share certain principles and beliefs. The perspectives shared by indigenous activists include a strong belief in the rights of citizens to participate in making environmental decisions; a general distrust of government based on direct experiences with public officials and agencies; a basic belief that human health—rather than aesthetics, wilderness preservation, or other issues—is the primary concern of their community organization; a skepticism about science and industry; and a prevalent belief that economic growth is not necessarily good and does not benefit everyone equally. Likewise, the groups discussed here have achieved successes that are comparable to those of other grassroots environmental organizations.[27]

Challenging Traditional Definitions of Political Activism

The traditional national environmental organizations—including the "Dominant Dozen" and the "Group of 10"[28]—have been criticized for representing white, middle-class interests and employing, primarily, white males in their leadership.[29] These organizations focus a majority of their efforts on national legislation or litigation, and measure their success through changes in national public policies.[30] Whereas many of these organizations have regional offices and local chapters, most of their resources are allocated to their national offices, usually in Washington, D.C. Finally, a high proportion of their professional staffs consist of scientists and attorneys who focus their energies on litigation, legislation, and scientific activities.[31] Many of the national organizations recognize that grassroots groups form to fight localized health and quality-of-life issues. Simultaneously, the national groups believe that these grassroots organizations diffuse their focus and, hence, their efficacy[32] (see Longo, next chapter).

"This Is My Life!"

The strategies used by each indigenous woman include teaching community members to help themselves, providing advocacy, forming coalitions, and sharing information. The study participants defined themselves, their lives, and their work in terms of their relationship to their families and community. Each woman identifies strongly with her particular ethnic group. Cultural maintenance and diversity are important to these leaders (our findings reflect the discussion by Robyn and Camacho, this volume). That is, their cultures reflect and reinforce their people's dependence on and respect for the natural environment. Their spirituality, rituals, customs, and traditional agrarian practices emphasize the importance of environmental sustainability and a harmonious coexistence with nature.

These leaders, however, do not identify themselves as "ecofeminists," since many reject the category "feminist." The traditional functions of women in their cultures as life givers, caregivers, and teachers affirm the centrality of the feminine and, in the case of American Indian women, their essential spiritual role. Furthermore, many of the women interviewed refuse to place the value of animals or nature above that of human beings. Instead, they stress the need for maintaining a balance between human needs and those of other living beings. Their approach to the environment is one of pragmatic yet respectful conservation. While indigenous women of Spanish descent describe their people's role as environmental stewards, Native American women express a spiritual connectedness that incorporates yet transcends the notion of stewardship. Indeed, a study of their traditional beliefs and practices illustrates how integral the extrahuman world is to their lives. Nature instructs and provides, and it also weaves individuals and their communities together, supplying sustenance and direction for living.

Following their sense of interconnectedness with nature and their human community, there is no distinct separation between their personal identity and their work identities. Each indigenous leader indicated that she could not imagine herself doing anything else; her work is her life. Nearly all of the women admitted, albeit reluctantly, to being political actors in their communities. Their role as mother, wife, Chicana or Indian, professional, or community member superseded their identification as a viable political activist. This finding is remarkably similar to Carol Hardy-Fanta's study of Latina organizers in Boston.[33]

In the past, political scientists studying political behavior tended to ignore issues concerning women and people of color, or regarded these as "special topics"—in effect, marginalizing these areas within the discipline. Working women's activism has spread beyond issues of reproduction (the domestic sphere) into the productive sphere (the workplace and the larger community), raising the political consciousness of its participants. By obtaining support at the grassroots level and adopting active leadership roles in their communities, women assert their position in the policymaking process. Hence, we are able to gain a richer understanding of women's political behavior when traditional definitions of political participation are relaxed to include grassroots action directed at affecting environmental, economic, and social conditions and influencing public policy.[34]

Indigenous grassroots environmental justice groups in New Mexico will continue to demand justice, winning some battles and losing others. Our findings do not substantiate the notion that grassroots environmental activities are trivial, marginalized, parochial, or fail to go beyond a single issue. Rather, our interview data illustrate that the twelve indigenous leaders make substantial contributions to the improvement of the quality of life, health, and environment in their communities. As Freudenberg and Steinsapir observed:

> By raising health concerns and by linking environmental issues to struggles for social justice and equity, the grassroots environmental movement has created the potential for a cross-class movement with a broader agenda, more diverse constituencies and a more radical critique of contemporary society than that of the national environmental organizations.[35]

Admittedly, these research discoveries are limited because of the small number of interviews, the inherently localized focus, and the inability to generalize to the larger population of environmental justice organizations. The findings, however, based on rich field research data, provide a rare opportunity to develop an understanding of Native American women and Chicana organizers and the strategies they use in their struggle to secure environmental justice.

Appendix A

INTERVIEW GUIDE

Date:
Alias:
Length of interview:
Permission to tape record interview:
Guaranty of anonymity given:
Permission to use data from interview for research:
Organization's name:
Organization's address:
Organization's phone number:
Respondent's name:
Title within organization:
Length of time in this position:
Length of time with organization:

Demographics

1.0 What is your highest level of formal education?
1.1 Do you belong to a church or religious group? If so, which one(s)?
1.2 Are you married or currently involved in a committed live-in relationship?

1.3 What is your husband's (partner's) occupation?
1.4 Do you have any minor children (under eighteen) living with you? If so, how many?
1.5 Who has primary responsibility for housekeeping in your home?
1.6 How are household chores divided?
1.7 Roughly how much of the housekeeping are you responsible for (percentage)?
1.8 How old were you on your last birthday?
1.9 What was your total household income last year?

Self identification

2.0 What language do you feel most comfortable speaking?
2.1 What language(s) did you grow up speaking?
2.2 What language do you use most regularly?
2.3 What is your race or ethnic background?
2.4 What is your primary occupation?
2.5 Are you a primary income earner in your household?
2.6 How would you describe yourself?

Organizational goals/operations

3.0 Please describe your organization's purpose or mission.
3.1 What are your primary activities or duties?
3.2 Please describe your role within the organization.
3.3 What are your goals for the organization?
3.4 How does your organization do its business?
3.5 What are the various strategies your organization uses?
3.6 What tactics are used to reach your organization's goals?

Organizational dynamics

4.0 Why did you become involved in the organization?
4.1 How did you become involved?
4.2 How are decisions made?
4.3 How do you influence decision making within your organization?
4.4 Who makes the final decision?
4.5 Are men and women equally involved in decision making? Or is one group more influential?
4.6 Compared with other leaders in the organization, are you more or less involved?

Community involvement

5.0 What role do you play in the community? Please explain.
5.1 Have you had an impact within your community? How?
5.2 Have your efforts led to change? Explain.
5.3 Has your involvement in the community affected you? How? Has your involvement affected your relationship with:
5.4 Your husband (partner)? How?
5.5 Other family members? Who? How?
5.6 Friends? How?
5.7 Other members of the community? How?
5.8 What gets in the way of your community work?
5.9 What helps you with your community work?

Political beliefs

6.0 When did you first become involved in issues outside of your home? Why?
6.1 Do you consider yourself political? (If yes, ask A, B, C/If no, ask X, Y)
6.1A What political activities are you involved in?
6.1.B Of the people closest to you, who do you consider political? Why?
6.1.C What do they do politically?
6.1.X What is politics for you?
6.1.Y Are any of your family active in community issues? Who? How?
6.2 Are you registered to vote?
6.3 Do you vote regularly?
6.4 Do you identify with a political party? If so, which one?
6.5 Are you a feminist? Please explain.
6.6 Are you an environmentalist? Explain.
6.7 Are you a political activist? Explain.
6.8 Are you a community organizer? Explain.
7.0 Is there anything else you'd like to tell me about your organization, your involvement in the community, or yourself?
8.0 Last of all, which other Chicana and Indian women—who are involved in environmental issues—would you suggest I contact to interview? I need their organizations and phone numbers as well as their names.

Thank you very much.

Appendix B

Organizational Attributes

	Scope	*Goals*	*Strategies*	*Impact*
Carmen	Regional	1	B,C,D1,D2,I,L1,L2,N,P,T1	6,7,8,9,10,Z
Dalia	Local	2	A,B,D2,F,H,I,L1,L2,N,O,P, S,T1,T2	6,7,10,Z
Donna	Local	2	A,C,D1,D2,H,I,M,N,T1	6,10
Felicia	Regional	3	B,C,D1,E,L1,I,N,PS,T1,T2	6,7,8,10,Z
Jackie	National*	1	B,C,D1,D2,N,P,S,T1,T2	6,7,9,10,Z
Juana	Regional	4	A,AV,B,C,D1,D2,E,H,L1,L2, I,M,N,O,P,R,ST1,T2,V	6,7,8,10,Z
Leslie	International*	5	B,C,D1,D2,E,N,O,P,S,T1, T2,X	6,7,9,10
Linda	International*	5	B,C,D1,D2,L1,L2,I,N,O,P,S, T1,T2,X	6,7,8,9,10
Monica	Local	2	AV,C,H,L1,L2,I,M,N,O,P, S,T1	6,8,10
Rose	Statewide	1	B,C,D1,D2,H,L1,L2,I,N,O, P,R,S,T1,T2	6,7,10,Z
Susana	Statewide	3	B,C,D1,D2,E,H,L1,L2,I,M, N,O,P,T1,T2	6,7,10,Z
Terri	Local	4	B,C,D1,D2,E,H,L1,L2,I,M, N,O,P,R,S,T1,T2,V,X	6,7,8,9,10,Z

*Organizations working with Indian nations and other indigenous groups in the United States and in North and Central America.

GOALS LEGEND

1 To provide technical assistance and information to communities interested in solving their environmental problems and developing environmental programs.

2 To advocate community environmental health, to provide reparation to victims of toxic poisoning, and to remedy social and economic injustice in the community.

3 To advocate community participation and influence in environmental policymaking.

4 To empower community members to achieve racial and gender equality, and social, environmental, and economic justice.

5 To advocate land, water, environmental, religious, and cultural rights of indigenous peoples.

Strategies Legend

STRATEGIES TO EDUCATE AND INFORM

B Capacity building: establishing leadership training for community members to increase public participation in policymaking.
C Conducting workshops, training seminars, or conferences to educate and inform.
E Educating and mentoring youth to enter into politically influential professions.
P Producing publications, such as newsletters and brochures, for wide distribution.
V Conducting voter registration drives or sponsoring political debates.

COMMUNITY RELATIONS STRATEGIES

A Gaining advocacy, endorsement, or sponsorship from politicians or celebrities.
AV Producing audiovisual presentations for public viewing.
H Holding well-publicized public hearings to inform and garner public support.
M Effectively using free media—radio, television, or newspapers—to garner public support.
R Organizing rallies, demonstrations, or boycotts.

PARTNERSHIP STRATEGIES

F Forming alliances with business and industry to invest in the community.
N Networking to exchange information, to share resources, or to develop strategies.
O Coalition building with friendly civic, social, labor, or professional organizations or religious groups.

STRATEGIES TO DEVELOP RESOURCES

D1 Develop reliable, credible sources of printed information based on accurate research.

S Fund-raising to gain financial support for operations and projects.
T1 Providing or acquiring technical assistance or expertise.
T2 Employing technical or legal experts as witnesses, advocates, or spokespersons.
X Repatriating land, cultural, and/or religious artifacts for preservation of community heritage.

STRATEGIES USING THE SYSTEM

D2 Reviewing official documents, and monitoring for regulatory enforcement and compliance.
L1 Lobbying to affect legislation at the local, state, tribal, or national levels.
L2 Using litigation to enforce compliance of laws or to win rewards for damages.
I Influencing local, state, tribal, and/or national political or bureaucratic officials.

IMPACT LEGEND

6 Increased community participation and galvanized community members.
7 Influenced public policy, passage of legislation, or enforcement of regulations.
8 Won litigation or legal settlement(s) affecting community resources or rights.
9 Preserved or repatriated land, or cultural or religious artifacts, or resources.
10 Acquired funding and/or official sanction to establish programs or implement projects.
Z Effected community environmental cleanup.

Notes

1. Sherry Cable and Charles Cable, *Environmental Problems, Grassroots Solutions: The Politics of Grassroots Environmental Conflict* (New York: St. Martin's Press, 1995); and Sherry Cable and Michael Benson, "Acting Locally: Environmental Injustice and the Emergence of Grass-Roots

Environmental Organizations," *Social Problems* 40, no. 4 (November 1993), 464.

2. Cable and Benson, "Acting Locally," 464.

3. Michael Edelstein, *Contaminated Communities: The Social and Psychological Impacts of Residential Toxic Exposure* (Boulder, Colo.: Westview, 1988); Nicholas Freudenberg, *Not in Our Backyards! Community Action for Health and the Environment* (New York: Monthly Review Press, 1984); and Benjamin A. Goldman, *The Truth about Where You Live: Atlas for Action on Toxins and Mortality* (New York: Time Books/Random House, 1991).

4. We will refer to members of nonwhite ethnic or racial minorities as *indigenous* or *people of color* because scholars using the minority model and respondents in this study use the term to refer to "Third World," Black, and colonized people in the United States. Bob Blauner, "Colonized and Immigrant Minorities," in *From Different Shores: Perspectives on Race and Ethnicity in America,* ed. Ronald Takaki (New York: Oxford University Press, 1987). The term *indigenous (people)* demonstrates both respect for and a solidarity with nonwhites in pre- and newly industrialized countries, and includes individuals of African and Asian descent as well as other nonwhite indigenous peoples. Dorceta Taylor, "Can the Environmental Movement Attract and Maintain the Support of Minorities?" in *The Proceedings of the Michigan Conference on Race and the Incidence of Environmental Hazards,* ed. Bunyan Bryant and Paul Mohai (Ann Arbor: School of Natural Resources, University of Michigan, 1990).

5. Robert D. Bullard, "Environmental Blackmail in Minority Communities," in Bryant and Mohai, The Proceedings of the Michigan Conference on Race and the Incidence of Environmental Hazards, 60–75; Taylor, "Can the Environmental Movement Attract and Maintain Support?"

6. Cable and Cable, *Environmental Problems,* 3.

7. Ibid., 104.

8. Ibid., 107.

9. Human Environment Center, *Mission and Activities Statement* (Washington, D.C.: Human Environment Center, 1991).

10. Bullard, "Environmental Blackmail," 60–75.

11. Taylor, "Can the Environmental Movement Attract and Maintain Support?" 39.

12. According to one source, "Air pollution in inner-city neighborhoods can be found at levels up to five times greater than those found in suburban areas. Urban areas, in general, have 'dirtier air and drinking

water, more wastewater and solid waste problems, and greater exposure to lead and other heavy metals than non-urban areas.'" Bullard, "Environmental Blackmail," 72.

13. Ann Bookman and Sandra Morgen, eds., *Women and the Politics of Empowerment* (Philadelphia: Temple University Press, 1988); and Marianne Githens, "The Elusive Paradigm Gender, Politics, and Political Behavior: The State of the Art," in *Political Science: The State of the Discipline,* ed. Ada Finifeter (Washington, D.C.: American Political Science Association, 1983), 471–99.

14. Sandra Morgen, "It's the Whole Power of the City against Us! The Development of Political Consciousness in a Women's Health Care Coalition," in Bookman and Morgen, *Women and the Politics of Empowerment,* 111–12. See also Bookman and Morgen, *Women and the Politics of Empowerment;* Irene Dabrowski, "Working-Class Women and Civic Action: A Case of an Innovative Community Role," *Policy Studies Journal* 11, no. 3 (1983): 427–35; Christine Marie Sierra and Adaljiza Sosa-Riddell, "Chicanas as Political Actors: Rare Literature, Complex Practice," *National Political Science Review* 4 (1994): 207–17; and Taylor, "Can the Environmental Movement Attract and Maintain Support?"

15. Karen Warren, ed., *Ecological Feminist Philosophies* (Bloomington: Indiana University Press, 1996).

16. In social research, we refer to assured confidentiality, which is the promise that the real names of the persons, places, and so forth will not be used in the research report or that they will be substituted with pseudonyms. John Lofland and Lyn H. Lofland, *Analyzing Social Settings: A Guide to Qualitative Observation and Analysis,* 3d ed. (Belmont, Calif.: Wadsworth Publishing, 1995), 43.

17. This is especially true in feminist research, where womens' stories (or personal narratives) can be used to systematically interpret the political models these women offer. Eloise Buker, "Storytelling Power: Personal Narratives and Political Analysis," *Women and Politics* 7, no. 3 (1987): 29.

18. Lewis A. Dexter, *Elite and Specialized Interviewing* (Evanston, Ill.: Northwestern University Press, 1970).

19. Grant McCraken, *The Long Interview* (Newbury Park, Calif.: Sage, 1988), 16.

20. Following the basic instructions for conducting field studies. Lofland and Lofland, *Analyzing Social Settings,* 1.

21. Lois Gibb, "Women and Burnout," in *Women and Burnout Fact Pack* (Arlington, Va.: Citizen's Clearinghouse for Hazardous Wastes, n.d.).

22. Bookman and Morgen, *Women and the Politics of Empowerment;* bell hooks, *Ain't I a Woman: Black Women and Feminism* (Boston: South End Press, 1981).

23. "Terri," personal interview, 25 November 1991.

24. Helen M. Ingram and Dean E. Mann, "Interest Groups and Environmental Policy," in *Environmental Politics and Policy: Theories and Evidence,* ed. James P. Lester (Durham, N.C.: Duke University Press, 1989), 157.

25. Frank Baumgartner and Bryan Jones, *Agendas and Instability in American Politics* (Chicago: University of Chicago Press, 1993); Anthony Downs, *An Economic Theory of Democracy* (New York: Harper and Row, 1957); Charles O. Jones, *An Introduction to the Study of Public Policy* (Monterey, Calif.: Brooks-Cole Publishing, 1984); and Paul Sabatier and Hank Jenkins-Smith, eds., *Policy Change and Learning: An Advocacy Coalition Approach* (Boulder, Colo.: Westview Press, 1993).

26. Nicholas Freudenberg and Carol Steinsapir, "Not in Our Backyards: The Grassroots Environmental Movement," *Society and Natural Resources* 4 (1991): 235–45.

27. Freudenberg and Steinsapir, "Not in Our Backyards."

28. The "Dominant Dozen" are the Sierra Club, National Audubon Society, National Parks and Conservation Association, Izaak Walton League, Wilderness Society, National Wildlife Federation, Defenders of Wildlife, Environmental Defense Fund, Friends of the Earth, Natural Resources Defense Council, Environmental Action, and Environmental Policy Institute. A subset of these compose the "Group of 10." Cable and Cable, *Environmental Problems,* 71.

29. Ingram and Mann, "Interest Groups," 135–57; and Freudenberg and Steinsapir, "Not in Our Backyards," 240.

30. Cable and Cable, *Environmental Problems;* and Freudenberg and Steinsapir, "Not in Our Backyards."

31. Freudenberg and Steinsapir, "Not in Our Backyards," 239–40.

32. Ingram and Mann, "Interest Groups," 150–55.

33. Carol Hardy-Fanta, *Latina Politics, Latino Politics* (Philadelphia: Temple University Press, 1993).

34. Bookman and Morgen, "Women and the Politics of Empowerment"; and Githens, "Elusive Paradigm."

35. Freudenberg and Steinsapir, "Not in Our Backyards," 242–43.

Peter J. Longo

ENVIRONMENTAL INJUSTICES AND TRADITIONAL ENVIRONMENTAL ORGANIZATIONS Potential for Coalition Building

David Truman, J. W. Peltason, and others have suggested that groups often pressure courts to change or alter public policies.[1] Karen Orren noted that in the 1970s standing problems—the requirement that a plaintiff must show that he or she has suffered a legal injury (the plaintiff must have personally experienced the injury) to a protected interest or right, and that other opportunities for defending the claim have been exhausted—at the federal court level were greatly reduced and "interest groups, frustrated with Congress, made an end run to the courts."[2] But where exactly do environmental interest groups do their running? Groups like the National Wildlife Federation and the Sierra Club are well known for their impressive success in protecting the environment in the "wilderness." Certainly, protecting the wilderness in pristine settings is of great importance, but environmental concerns are also readily found in urban areas. Indeed, urban environmental threats often touch on the area of environmental racism.

Many urban citizens of color and members of low-income groups subjected to environmental injustices could benefit from an alliance with the highly influential Sierra Club and National Wildlife Federation. Such alliances would be of great worth, given the amount of environmental litigation and resulting policy. Theodore Lowi argued that lawmakers have given up on making laws.[3] U.S. legislators, faced with politically volatile problems and pressured by deeply committed groups, have willingly forfeited and delegated their lawmaking powers to the courts.[4] David Rosenbloom viewed the process as the "judicialization" of public policy.[5] Undoubtedly, in democratic regimes like the United States, it is the role of the legislative branch to provide meaning to environmental issues. Because of the process described, this is typically not the case.

Zachary Smith suggests that many solutions exist for environmental problems, but the policymaking process prevents their implementation.[6] As Donald Horowitz argued, this process is different in the judicial arena, as "more often than not, the judicial remedy has a directness, a concreteness, and a lack of equivocation notably absent in schemes that emerge from the political process."[7] The values related to justice need to be heard and the courts are an important venue for them to be debated. It would be a mistake to proceed under a romantic assumption that the courts provide nothing but justice. John Rawls makes the case that justice is the first virtue of social institutions, but that many disagree as to the definition of justice.[8] In more general terms, Walter Murphy, Lawrence Baum, and others clearly illustrate the limitations of the courts in providing universal meaning.[9] Nonetheless, the study of court decisions provides valuable instruction to the student of environmental politics and policy.

The Sierra Club and National Wildlife Federation's experiences in the judicial process would be of great importance to less-experienced, urban-based environmental groups. This chapter analyzes the landmark Supreme Court cases of *Sierra Club v Morton, Lujan v National Wildlife Federation,* and *Ruckelhaus v Sierra Club.* These cases offer "cues" that could provide meaningful insight and direction to those citizens affected by the practices leading to conditions of environmental injustice.

The Need: Environmental Injustices

The notion of environmental injustice is not new nor is it one conjured up by academics in ivory towers. Other articles in this anthology have devoted considerable attention to the political process model, which points to the importance of established organizations and their leaders in denouncing environmental injustices. Practices leading to environmental injustices persist; White's empirical assessment (this volume) describes the continued injustices endured by people of color and low-income groups. Certainly, such practices are somewhat problematic because to some people of color, such as various Native American groups, contracting for dumping sites has become a means of economic survival.[10] Regardless, the evidence is clear—people of color and low-income groups live in neighborhoods that were former industrial dumping grounds or are still active dumping sites.

Such urban dumping grounds have not been the typical focus of National Wildlife Federation or Sierra Club activity. As noted by Robert Gottlieb, the Sierra Club "divorced themselves from the issues and concerns of radical urban and industrial movements of the Depression years."[11] It was also reported that, "the early criticism, at best, described the environmental movement as irrelevant and ignoring the needs of minority and economically disadvantaged movements."[12] Additionally, the Sierra Club and National Wildlife Federation are perceived to have set the national agenda, giving unparalleled legitimacy to these mainstream efforts. As Robert Mitchell, Angela Mertig, and Riley Dunlap noted, "When most people think about the environmental movement they are likely to think first of large national environmental organizations such as the Sierra Club, the National Audubon Society, or the National Wildlife Federation."[13] There is a perception, thanks to the success of national groups, that environmentalism is only needed in remote areas. Further, there is a perception that local groups and national groups are in competition for attention. But as Sierra Club insider Michael McCloskey observed, competition needs to be harnessed and "where the problem lies is not in this healthy competition, but in the absence of healthy interaction."[14] What is needed, in other words, are clear lines of communication between local and national groups.

Consequently, for intentional or unintentional reasons or motivations, the Sierra Club and National Wildlife Federation had little to do with spotlighting urban environmental problems. Although each incident of environmental injustice has unique features, a telling example of this condition was reported in a lead editorial in the *Rocky Mountain News.* An all-too-familiar account follows:

> Lorraine Granado walked into the offices of the Cross Community Coalition recently to discuss the environmental hazards surrounding communities in North Denver. Suddenly an odor permeated the room. . . . [D]epending on the wind currents, Granado says the smell could be from the Ralston Purina Consumers Products factory, a sewage treatment plant, a lamb rendering plant, nearby refineries of all four. . . . "I have learned to distinguish, believe me. We're completely surrounded."[15]

Of particular interest to this chapter is the fact that existing grassroots efforts are underway to ameliorate the effects of environmental injustices. For instance, the statewide Colorado People's Environmental and Economic Network is currently attempting to assist affected people of color and low-income groups. This type of grassroots organization would quite naturally benefit from Sierra Club recognition and involvement; conversely, the Sierra Club would be able to utilize the cause of such groups to broaden its scope and impact to a wider audience of citizen activists.

Another account of the many facets of environmental injustice was provided in a 1994 *New York Times* article, "For the Poor, a Legal Assist in the Cleanup of Pollution," which reported:

> For years, Olivia Barros has fought illegal dumping in her neighborhood—even in her own backyard. The campaign that she and her neighbors waged in the Dorchester section here resulted in the city's carting away five truckloads of trash in 1988, but the garbage always returned.[16]

Obviously, illegal dumping in neighborhoods where the poor and people of color live is newsworthy on its face. Of more significance is the fact that Barros was assisted by the Massachusetts Environmental Justice Network. In this case, the Boston-area lawyers were willing to take on the cause of environmental injustice. But environmental law is highly complex, demanding seasoned environmental litigants like the Sierra Club and National Wildlife Federation.

The call for such partnerships is not new, as reported in the *San Diego Union-Tribune* story, "Minorities Outraged at Burden of Toxics: 'Environmental Equity' Demanded across the U.S."[17] The article details how the Sierra Club is gradually becoming a partner in the fight against environmental injustices. The *Union-Tribune* detailed that in the past, the Sierra Club has focused a significant amount of attention on park preservation and coastlines, but the organization has been very sensitive to environmental equity and racism. Further, as James Colopy noted, new alliances are being forged between traditional racism.[18] As we will see in the following landmark cases, partnerships between citizens suffering from environmental injustices and these two powerful national organizations are mutually beneficial, if not essential. Mainstream environmental organizations are increasingly supporting grassroots groups in the form of technical advice, expert testimony, direct financial assistance, fundraising, research, and legal assistance.

Sierra Club v. Morton

While the Sierra Club has fought many important judicial battles throughout its political history, the issue of *standing*, or legal injury, is an obstacle to success, as exemplified in *Sierra Club v. Morton.*[19] In this 1972 case, the Sierra Club opposed the federal approval of an extensive ski development in the Mineral King Valley in California's Sequoia National Park on the basis of the aesthetic concerns of its membership. The majority held that the Sierra Club lacked standing to assert its claim. Specifically, the Supreme Court held that

> aesthetic and environmental well-being, like economic well-being, are important ingredients of the quality of life in our society, and the fact that peculiar environmental interests are shared by many rather than the few does not make them less deserving of legal protection through the judicial process. But the injury in fact requires more than injury to a cognizable interest. It requires that the party seeking review be among himself the injured. . . . Nowhere in the pleadings or affidavits did the Club state that its members use Mineral King for any purpose, much less that they use it in any way that would be significantly affected by the proposed actions of the respondents.[20]

In a dissenting opinion, Justice William Douglas offered the following as a jurisprudence of standing:

> The critical question of "standing" would be simplified and also put neatly into focus if we fashioned a federal rule that allowed environmental issues to be litigated before federal agencies or in the name of the inanimate object about to be despoiled, defaced, or be invaded by roads and bulldozers and where the injury is the subject of public outrage. Contemporary public concern for protecting nature's ecological equilibrium should lead to the conferral of standing upon objects to sue for their own preservation.[21]

If ever widely adopted, Douglas's dissent would certainly benefit the cause of environmentalism, for an inanimate object could claim that it had personally experienced a legal injury! But unfortunately this jurisprudence is not grounded in recent political realities.

Judicial pragmatism, on the other hand, indicates that race is a cause of environmental deprivation. Environmental injustice, then, provides a different group of litigants and, in turn, a ground for standing in that these litigants are adversely affected by environmental atrocities. Accordingly, the standing issue should not be an obstacle in cases involving environmental injustice and the Sierra Club would be a welcome partner for those who suffer harm from its effect. That is, the Sierra Club, when joined with victims of environmental injustice, can present a united case that will not be muffled by the standing issue.

Additionally, Douglas's plea for a federal standing law would be needless in light of the local nature of environmental injustice. Standing would hinge on local or state laws. Indeed, as Peter Reich explained, "Recently, in the wake of federal entitlement cutbacks, scholars have begun to argue that state law should be interpreted liberally to address the specific needs of disadvantaged groups. In the absence of viable federal relief, state precedent and statutory provisions can be used to remedy environmental racism."[22] Given the temperament of the 104th Congress, it appears that Reich's observations persist. Citizens affronted by environmental injustices must, out of practical necessity, do battle in the state arena. The Sierra Club and National Wildlife Federation can greatly assist aggrieved citizens. At the same time, the Sierra Club and National Wildlife Federation will not only reduce their greatest obstacle to judicial success, standing, but they can broaden their environmental platform, likely resulting in a growth in membership as well.

Lujan v. National Wildlife Federation

This same vein of thought is illustrated in *Lujan v. National Wildlife Federation.*[23] In this 1989 case, the National Wildlife Federation claimed that the director of the U.S. Bureau of Land Management had violated the Federal Land Policy Act of 1976 and the National Environmental Policy Act of 1969. Specifically, the National Wildlife Federation "averred that the reclassification of some withdrawn lands and the return of others would open the lands up to mining activities, thereby destroying their natural beauty."[24] Once again, an environmental group was arguing that its members were deprived of enjoying the beauty of a wildlife setting.

In attempting to establish standing, the National Wildlife Federation provided the following affidavit:

My recreational use and aesthetic enjoyment of federal lands, particularly those in the vicinity of South Pass-Green Mountain, Wyoming have been and continue to be adversely affected by the unlawful actions of the Bureau and the Department. In particular, the South-Pass-Green Mountain area of Wyoming has been opened to the staking of mining claims which threatens the aesthetic beauty and wildlife habitat potential of these lands.[25]

The affidavit is much different than specific accounts of environmental injustice in that standing is not a question of immediate survival should the courts disallow judicial treatment of the issue. Supreme Court Justice Antonin Scalia, writing for the majority in *Lujan,* failed to see the establishment of standing by the National Wildlife Federation and ruled that standing is "not satisfied by averments which state only that one of the respondent's members uses unspecified portions of an immense tract of territory, on some portion of which mining activity has occurred or probably will occur by virtue of government action."[26] The majority opinion requires specific facts to support the environmental claim and the "land withdrawal review program is not an identifiable action or event."[27] Rather, specificity of harm is required. Victims of environmental injustice could quite conveniently provide the National Wildlife Federation with the needed specificity, while in turn, the National Wildlife Federation could supply the victims of environmental injustice with legal and organizational expertise.

The *Lujan* case is important for yet another reason. If we are to believe the dictum of Scalia, it is crucial that environmentalists pay careful attention to both the legislative and judicial processes.

The case-by-case approach that this requires is understandably frustrating to an organization such as respondent, which has as its objective across-the-board protection of our nation's wildlife and the streams and forests that support it. But this is the traditional, and, remains the normal, mode of operation of the courts. Except where Congress explicitly provides for our correction of the administrative process at a higher level of generality, we intervene in the administration of the laws only when, and to the extent that, a specific "final agency action" has an actual and immediately threatened effect. . . . Until confided to us, however, more sweeping actions are for the other branches.[28]

Scalia's statement captures the jurisprudence of a conservative court as well as the political sentiments of the 1990s' electorate and elected. Environmental groups must be prepared to fight their legal battles case by case at both state and federal levels. The comprehensive struggle ought to include scenarios like the *Lujan* case in addition to the many cases of environmental injustice. Yet, as discussed before, it is the latter cases that will allow groups to break down the legal obstacle of standing. Moreover, Scalia implies that the Court would be receptive to legislative directives. When citizens affected by environmental injustice pressure local lawmakers to provide the framework for environmental relief, the dictum suggests that relief should follow. It is clear that adherents of Scalia's environmental jurisprudence will not allow for an injury based on speculation nor for relief not anticipated by the legislative branch.

Ruckelhaus v. Sierra Club

The final case considered is *Ruckelhaus v. Sierra Club.*[29] This 1982 case pertains to the practical and often controversial issue of attorney fees. Judicial activism is not without expense. The cost of litigation often makes legal action prohibitive. The *Ruckelhaus* case serves as a reminder that attorney fees are a reality. The facts of this case reveal that the Sierra Club unsuccessfully challenged EPA standards limiting the emission of sulfur dioxide by coal-burning power plants.[30]

Citing section 307(f) of the Clean Air Act (the litigation costs section), Supreme Court Justice William Rehnquist wrote for the majority: "We conclude that the language of the section, read in light of the historic principles of fee-shifting in this and other countries, requires the conclusion that some successes on the merits be obtained before a party becomes eligible for a fee award under section 307(f)."[31] In other words, the Sierra Club needed to successfully litigate at least a portion of the claim before compensation for legal costs was justified.

Rachel Godsil, in a comprehensive work, revealed many obstacles to grassroots litigation by those suffering from environmental injustice, including the lack of financial resources.[32] Combined with the complexity (legal expertise is a must) and time-consuming nature (case-by-case approach) of the judicial process, a lack of financial resources presents a formidable obstacle for grassroots groups to successful legal insurgency. As the political process model indicates, established organizations normally have the kinds of resources that often extremely limited-income grassroots organizations lack. Established environmental groups could assist people of color and low-income groups by taking on the financial burden of costly litigation.

Conclusion

Perhaps it is too much to conclude that powerful national environmental interest groups and groups formed to combat environmental injustices can mutually benefit from new alliances. The observations made by Gary Machlis about the natural tension between local and national groups offer the following dismal possibility:

> Since it is confrontation and struggle that energize local groups the two areas of action will be separated by a lack of common purpose, different styles of action, and even different motivations. The distance between the office of the executive director of a national conservation group and the living room meeting place of a local group of angry citizens can be very, very far. Hence, the central irony of conservation in the democratic regime may be that sometimes conservation groups rob power instead of give power, and thus resemble the architects of dominion and environmental disregard.[33]

Yet such a possibility need not necessarily come true. Camacho (earlier in this volume) has discussed the factors for successful insurgency. Sandweiss (also this volume) has utilized the framework of social movement theory to make a similar argument to the one advanced here. This chapter has also suggested that people of color and low-income groups can: (1) provide mainstream environmental interests with a powerful and immediate electoral advantage at the local level; (2) enhance the national reputation, and thus legitimacy, of mainstream groups; and (3) add to the membership rolls of mainstream groups.

As Scalia stated in *Lujan,* the court is waiting for legislative direction to address national environmental issues. Electoral advantages may produce this direction. Linda Blank commented about the need for grassroots support for the on-again, off-again Environmental Justice Act of 1992.[34] More precisely, Blank suggests that community participation is an essential component of the environmental civil rights movement and is particularly critical to the effectiveness of the Environmental Justice Act. Stated differently, enforcement might be greatly enhanced through an alliance. Further, people of color and low-income groups provide a glimpse of a universal environmental picture. National environmental groups can make certain that the legislative message is not lost in the judicial arena. The proven legal savvy and resources of national groups are needed for successful litigation within local courts. While local courts might not seem important at first glance, they are often the first and last chance of relief for victims of environmental injustice, becoming true courts of last resort.

In the final analysis, alliances between national, mainstream environmental groups and local, indigenous groups formed to battle environmental injustices can directly bolster the democratic process. Indeed, as the focus of democratic power switches from the "shining city on the hill" to the fifty "star cities," the fluidity provided by links to all levels of government is gained by alliances between locally oriented groups and national organizations. Citizen participation and cooperation of this type not only fosters democracy, but can make environmental justice a real rather than an abstract concept.

Moreover, alliances challenge economic profit—the driving force that allows environmental injustice to persist. This is not to suggest an end to capitalism, for it is a highly productive economic system. One need only examine the economic strategies of the former Soviet Union, as one example, to assess the consequences of other economic theories. Rather, we need to be critical of the negative byproducts of capitalism. Lawrence Tribe argues for a changed "legal and constitutional framework for choice," a framework that allows us to rescue our perception of nature from the "conceptually oppressive sphere of human want and satisfaction."[35] Tribe's observation speaks to the negative effects of the U.S. "consumer society" as described by Camacho (later, this volume). "Quality of life" issues (Robyn and Camacho, also this volume) embraced by both mainstream and environmental justice movements must be adjoined—preservation and conservation of the environment can contribute to a healthy human condition. Accordingly, national, mainstream environmental groups need to inject *humanity* into their legal formulas. A concern with this ethical dimension raises the possibilities for environmental justice.

Notes

1. David D. Truman, *The Governmental Process, Political Interests, and Public Opinion* (New York: Knopf, 1951); and J. W. (Jack Walker) Peltason, *Federal Courts in the Political Process* (New York: Random House, 1955).
2. Karen Orren, "Standing to Sue: Interest Group Conflict in the Federal Courts," *American Political Science Review* 70 (1976): 723.
3. Theodore J. Lowi, *The End of Liberalism* (New York: Norton, 1969).
4. Douglas H. Shumavon and H. Kenneth Hibbelin, eds., *Administrative Discretion in Public Policy Implementation* (New York: Praeger, 1986).
5. David Rosenbloom, "Judicialization of Public Policy," *Administration and Policy Journal* 3 (1983).
6. Zachary A. Smith, *The Environmental Paradox* (Englewood Cliffs, N.J.: Prentice Hall, 1992).
7. Donald L. Horowitz, *The Courts and Social Policy* (Washington, D.C.: Brookings Institute, 1977), 11.
8. John Rawls, *A Theory of Justice* (Cambridge, Mass.: Belknap Press of Harvard University Press, 1971).
9. Walter F. Murphy, *Elements of Judicial Strategy* (Chicago: University of Chicago Press, 1964); and Lawrence Baum, *The Supreme Court,* 3d ed. (Washington, D.C.: CQ Press, 1989).
10. "A Winning Hand?" *National Journal,* 17 July 1993, 1796; and "NSP Finds Site for Nuclear Waste—At Least Temporarily," *Energy Daily,* 26 January 1996.
11. Robert Gottlieb, *Forcing the Spring: The Transformation of the Environmental Movement* (Washington, D.C.: Island Press, 1993), 74.
12. "Environmental Justice," *Recorder* (autumn 1994).
13. Robert Cameron Mitchell, Angela G. Mertig, and Riley E. Dunlap, "Twenty Years in Environmental Mobilization: Trends among National Environmental Organizations," in *American Environmentalism: The U.S. Environmental Movement,* Riley E. Dunlap and Angela G. Mertig (New York: Taylor and Francis, 1992), 11.
14. Michael McCloskey, "Twenty Years in the Environmental Move-

ment: An Insider's View," in Dunlap and Mertig, *American Environmentalism: The U.S. Environmental Movement,* 85.

15. "Fighting Environmental Racism," *Rocky Mountain News,* 21 November 1994, 4n.

16. "For the Poor, a Legal Assist in the Cleanup of Pollution," *New York Times,* 16 December 1994, 21.

17. "Minorities Outraged at Burden of Toxics: 'Environmental Equity' Demanded across the U.S.," *San Diego Union-Tribune,* TK, A3.

18. James H. Colopy, "The Road Less Travelled: Pursuing Environmental Justice through Title VI of the Civil Rights Act of 1964," *Stanford Environmental Law Journal* 13 (1994): 143.

19. *Sierra Club v Morton,* 405 US 727 (1972).

20. Ibid., 734–35.

21. Ibid., 742.

22. Peter L. Reich, "Greening the Ghetto: A Theory of Environmental Race Discrimination," *Kansas Law Review* 41 (1992): 300.

23. *Lujan v National Wildlife Federation,* 497 US 871 (1989).

24. Ibid., 879.

25. Ibid., 886.

26. Ibid., 889.

27. Ibid., 899.

28. Ibid., 864.

29. *Ruckelhaus v Sierra Club,* 463 US 680 (1982).

30. Ibid., 681.

31. Ibid., 682.

32. Rachel D. Godsil, "Remedying Environmental Racism," *Michigan Law Review* 90 (1991): 395–427.

33. Gary E. Machlis, "The Tension between Local and National Conservation Groups in the Democratic Regime," *Policy Review* 3 (1990): 278.

34. Linda Blank, "Seeking Solutions to Environmental Inequity: The Environmental Justice Act," *Environmental Law* 24 (1994): 1125.

35. Lawrence Tribe, "Ways Not to Think about Plastic Trees: New Foundations for Environmental Law," *Yale Law Journal* 83 (1974): 1338.

IV

Environmental Justice

Mary M. Timney

ENVIRONMENTAL INJUSTICES

Examples from Ohio

There is a growing movement for environmental justice across the United States as a result of several studies documenting the disproportionate siting of waste and polluting facilities in poor and minority neighborhoods.[1] Other chapters in this volume describe these studies in detail.

In this chapter, I argue that the term *environmental racism* is misleading, since it limits those affected to people of color. In Ohio, a high percentage of the poor are white Appalachians or blue-collar ethnics who live in the most polluted areas along with Blacks and other minorities. The term *environmental injustice,* on the other hand, describes the results of political and economic decisions that have disproportionate impacts on poor and minority populations. The critical issue is how state and city policymakers and public administrators have enforced pollution control laws and approved siting of new facilities in these areas.

Environmental justice is defined by the EPA as "the fair treatment of people of all races, cultures, and income with respect to the development, implementation, and enforcement of environmental laws, regulations, and policies."[2] Environmental injustice occurs when polluting facilities are disproportionately sited in neighborhoods or communities with a population of people of color or at economic disadvantage that also has little political power to influence the siting decision.

Critics of the above mentioned studies have cited problems with the definitions of environmental injustice, equity, or racism; the methodologies of the studies; and the assumption of the researchers that siting decisions have been a matter of policy as opposed to economics or market dynamics.[3]

Some studies have been criticized for correlating siting by zip code, geographic areas designated by the U.S. Postal Service that do not necessarily define minority/low-income neighborhoods. More recent studies have looked at census tracts, designed by the Census Bureau to have uniform socioeconomic characteristics.[4] Other studies have correlated emissions data from the Toxic Release Inventory (TRI)[5] by census tract.[6] They have not found strong correlations with race in the siting of facilities; rather, these studies seem to indicate a correlation with economic factors, particularly blue-collar job availability.

There is no evidence from the correlations alone of consciously racist decision making on the part of policymakers and industry, it is argued. Indeed, in many cases, it can be shown that the neighborhood grew up around the facility.

Critics also contend that the nuisance of pollution in low-income and minority communities is the trade-off that these residents make in order to provide for their families. As the saying goes in the old industrial neighborhoods, "Smoke in the skies means bread on the table." Christopher Boernor and Thomas Lambert propose that environmental justice can be achieved by providing some form of economic compensation to the communities for their disproportionate burden of the externalities of industrial production and waste disposal.[7]

Further, critics maintain that residential patterns are strongly linked to economic and cultural factors. Specifically, people live where the jobs are and the facilities represent economic opportunity for low-income communities willing to make the trade-off of pollution for jobs. Second, property values tend to be lower in areas of industrial development, providing housing opportunities for low-income people. Third, people want to live with others of their own culture. Consequently, we would expect to find higher concentrations of the poor and minorities in polluted areas where residential property is more affordable, especially if their families and friends also live there.

William Ryan calls arguments such as these "blaming the victim," because they overlook the reality of the limited choices that low-income people have in a society increasingly segregated by income.[8] They assume the mechanics of trickle-up opportunity: when incomes rise because of the economic opportunity presented by the polluting facility, then people of color and the poor will gain economically, allowing them to move away from the polluted areas. If they stay, it must be because they choose to live with their community. This argument overlooks the fact that as housing costs have risen since the late 1970s, the availability of low-income housing has declined everywhere. People of color and the poor increasingly have fewer housing choices than the affluent.

Moreover, it is difficult to measure whether polluting facilities have positive or negative effects on the economy of the impacted community. It is often claimed that poor communities welcome dirty industry and waste disposal facilities because of the jobs that they bring. If the siting of polluting facilities is an economic benefit for the community, then should we not expect to see some differential in the unemployment rate in the affected areas? Would unemployment rates decline as emissions increase? Setting aside for a moment the moral question of whether people should be expected to jeopardize their health and that of their children in order to provide for their families, is there a way to measure the true economic benefits of a polluting facility in comparison to the community's health and environmental costs?

We think it is irrelevant whether environmental injustice represents conscious racism, or classism, on the part of policymakers, either now or in the past. Attempting to prove intent is a fool's errand, particularly when there are so many variables in the mix. What truly matters is how the problems are addressed by policymakers and business interests in the present. In other words, who benefits from current siting and pollution control practices, and who pays the consequences? Further, how are these impacts affected by ongoing policies and enforcement practices?

Overview and Methodology of the Study

This chapter reports the results of a study conducted by the Ohio Environmental Council in the summer of 1993.[9] The study matched emissions data from the 1991 TRI with zip codes in Ohio's major cities: Cleveland, Cincinnati, Columbus, Toledo, Akron, Canton, Youngstown, and Dayton. The zip code areas with the highest emissions in each city were identified, and the sample was further reduced to those areas where emissions were at least 10 percent of the total emissions for the county in which the cities were located. We then examined the socioeconomic characteristics of the residents using 1990 census data. The characteristics selected were average income, percentage of residents over eighteen without a high school diploma (including those with less than a ninth grade education), percentage of families below the poverty level, percentage of people of color,[10] and percent unemployed, broken down further by race. The meaning of the findings was then assessed in relation to state environmental policies and enforcement practices.

We limited the same to the large cities in order to avoid diluting the data for racial minorities by including rural, largely white, areas of the state. Furthermore, the cities developed around big industrial facilities; if there are favorable economic effects from pollution, they are more likely to be seen in the cities than in rural areas, where the facilities are more scattered.

Certainly, there are weaknesses inherent in this methodology. The TRI is a measure of how much pollution is in an area, but not how toxic it is. Thus, smaller amounts of highly toxic materials can pose a greater health danger than much larger amounts of materials of lesser toxicity. Using gross data from the TRI can mislead, since the emissions may be more of a nuisance than a hazard. Second, zip code areas are not a sufficiently precise unit on which to base definitive judgments since they are geographic areas that provide the boundaries for individual post offices. They do not have the uniform socioeconomic characteristics of census tracts; thus, the data lack the power to demonstrate conclusively that environmental racism exists.

The purpose of this study, however, was not to prove that environmental racism or injustice exists; it is self-evident that the poor generally live in the most polluted areas of Ohio's cities, and further, that policymakers do not site polluting facilities in high-income neighborhoods and suburbs. Rather, the purpose was to identify some parameters of the problem in Ohio and to examine state and local policies that tend to perpetuate it.

Environmental Protection in Ohio

Ohio is a state of contrasts, with eight large industrial cities amid wide swaths of rural areas, mostly farming communities. It has ranked third among the states in toxic releases since the TRI was established in 1987, behind Texas and California.[11] The state's politics have long been dominated by industrial interests, which have restrained the implementation and enforcement of environmental protection laws. Ohio was a major opponent of 1990 congressional amendments to the Clean Air Act to reduce acid emissions, largely because electric utilities in the state burn Ohio coal and are big generators of sulfur dioxide and carbon dioxide. Ohio officials led the 1994 states' fight against unfunded federal mandates, essentially a rebellion against environmental protection laws. The state's population includes 12 percent minorities or people of color, 85 percent of whom are Black. It also has a sizable poor white population, primarily consisting of Appalachians from West Virginia and Kentucky. The old factory cities—Cleveland, Youngstown, Canton, and Akron—have significant populations of blue-collar workers of European ethnic backgrounds. Twenty-four percent of the adults over eighteen in Ohio do not have high school diplomas. The cities in Ohio have strong neighborhood identities, which may have a bearing on why people live where they do.

Politically, the state is relatively conservative even though, until recently, Democrats held both the governor's office and the state House of Representatives. In fall 1994, however, Republicans took control of all branches of the state government. The state has never been a leader in implementing environmental protection laws. It was one of the last in the nation to require inspection and maintenance of automobiles in noncompliance air basins and has dragged its feet in funding environmental cleanup projects. In the summer of 1993, the governor and Ohio Environmental Protection Agency (OEPA) approved the release of millions of gallons of acid mine drainage into a rural creek, despite the objections of the U.S. EPA, in what was clearly a violation of the Federal Water Pollution Control Act. The state argued that the 800 coal mining jobs that *might* be saved had more value than the watershed.

Ohio ranks somewhere in the low middle of states in enforcement of environmental protection laws. The OEPA, which has been underfunded and understaffed for a long time, tends to take a minimalist approach to enforcement, particularly where major industrial interests are concerned. Since George Voinovich became governor in 1991, state funding of the OEPA's budget has been continually reduced; budget cuts have been partially offset by increased fee income and some federal funding.

The governor has made it clear throughout his tenure that his main goals are to increase jobs in the private sector and reduce the size of government. Environmental policy must also achieve these goals. Since 1994, the governor and the Republican-majority legislature have worked to develop environmental regulations that limit liability of polluting companies and promote business development.

There has been some cleanup in the past twenty years—the Cuyahoga River in Cleveland no longer bursts into flames—nonetheless, state policies continue to tolerate high levels of pollution and environmental degradation throughout Ohio, particularly in poor, politically weak areas in cities and rural locations.

Findings of the Study

Eleven zip code areas were identified as having more than 10 percent of the total toxic releases within their county. Table 1 summarizes the educational and racial characteristics of these areas in comparison with statewide averages. Within the zip codes, the average income ($21,481) was about 50 percent of the statewide average income ($40,813); the number of residents without a high school education was about 40 percent greater than the state average, while the number of families below the poverty level was more than 75 percent higher than the state average. The number of people of color living in these areas was 120 percent higher than the state average.

Unemployment in these areas was generally higher than the state average (about 80 percent). White unemployment in the zip code areas was 54 percent higher than the state average white unemployment, but only 33 percent higher than the state average. Black unemployment was 60 percent higher than the state average Black unemployment and 180 percent higher than the average unemployment in the state.

Table 2 provides a breakdown of income, education, poverty level, and racial data within the zip codes along with total toxic emissions in thousands of pounds. Toxic chemicals are released into the air, water, and soil of the immediate areas surrounding the sources of emissions, and some are transferred to treatment facilities both within the state and outside of Ohio. Toxic chemicals are also imported into Ohio for disposal. This study identified total toxic releases in the zip code, regardless of destination.

Table 1. Demographic Characteristics of the State of Ohio and Eleven Ohio Zip Codes (all figures are percentages)

	State	*Eleven Zip Codes*	*Percentage of Total*
Without a high school diploma	24.0	33.0	138.0
Poverty level	12.5	22.0	176.0
People of color	12.0	26.5	221.0
Unemployment	6.6	10.8	182.0
White	5.7	8.7	154.0
Black	11.6	18.6	160.0

Note: White unemployment in the eleven zip code areas is 8.7/6.6 = 133 percent of average state unemployment, while black unemployment in these areas is 18.6/6.6 = 280 percent of average state unemployment.
Source: Bureau of the Census, 1990.

Toxic releases in the state of Ohio in 1991 totaled 2.5 billion pounds. Cuyahoga County, in which Cleveland is located, had 52 million pounds, followed by Canton in Stark County, an old manufacturing city, with 35 million pounds.[12] Hamilton County, home to Cincinnati, had 27 million pounds of toxic emissions and Franklin County, where the state capital of Columbus is located, had 21 million pounds.

The areas identified in table 2, for the most part, show higher poverty levels with generally more people of color than the statewide averages. The range in the zip codes is from 4 to 71 percent people of color. This illustrates the difficulty in trying to demonstrate environmental racism where the affected populations are already a significant minority of the total population. What is readily apparent from table 2 is that high toxic releases occur in neighborhoods where average incomes are quite low and significantly high percentages of the residents are undereducated. This indicates the strong possibility that siting policies may be skewed toward these areas because of the perceived political weakness of poor people, in addition to racism, on the part of industry and government.

Unemployment statistics for the eleven areas are shown in table 3. The rate of unemployment exceeds the state rate in all but two areas. Black unemployment exceeds the state rate of Black unemployment in all but three areas and, in all areas but one, the rate of white unemployment is significantly below that of Blacks.

Table 2. Toxic Emissions and Demographic Characteristics of Eleven Ohio Zip Codes (1991)

County/City Zip Codes	*Total County Emissions (000 lbs.)*	*Zip Code Emissions (000 lbs.)*	*Average Income*	*Percentage without High School Diploma*	*Percentage Poverty*	*Percentage People of Color*
Cuyahoga/ Cleveland	51,821					
• 44109		11,830	$24,286	37	11.0	12
Hamilton/ Cincinnati	26,841					
• 45217		5,794	22,746	32	9.5	38
• 45232		5,671	22,846	42	53.0	71
Franklin/ Columbus	20,762					
• 43207		2,854	27,389	42	14.5	25
• 43201		2,554	14,909	12	38.0	25
• 43215		2,322	18,590	28	23.0	24
Montgomery/ Dayton	11,234					
• 45404		1,190	16,425	44	27.0	12
Stark/ Canton	34,965					
• 44704		11,524	17,585	45	28.0	62
• 44706		4,834	21,238	31	11.0	6
Summit/ Akron	4,932					
• 44301		980	25,545	30	16.0	13
Mahoning/ Youngstown	4,859					
• 44515		2,415	24,733	22	8.0	4

Source: U.S. Environmental Protection Agency, Toxic Release Inventory, 1991; and Bureau of the Census, 1990.

There appears to be little positive relationship between levels of unemployment and levels of emissions within the areas. If the presence of emissions is an economic benefit for the residents, we might expect to see an inverse relationship between the unemployment rate and pounds of emissions—the higher the emissions, the lower the unemployment. The highest unemployment rate in this sample, however, is found in the area with the second highest emissions; low unemployment rates do not equate with high emissions or vice versa. These data do not necessarily refute the argument that pollut-

Table 3. Unemployment Rates in Eleven Ohio Zip Codes (1990)

Zip Code	*Emissions (000 lbs.)*	*Total Unemployment (percentage)*	*White Unemployment (percentage)*	*Black Unemployment (percentage)*
44109	11,830	10.9	9.2	17.5
45217	5,794	7.7	4.3	13.6
45232	5,671	18.8	7.5	26.8
43207	2,854	9.0	7.9	12.4
43201	2,554	8.4	7.4	12.4
43215	2,322	8.5	7.9	11.5
45404	1,190	12.8	10.7	42.2
44704	11,524	19.7	20.3	19.3
44706	4,859	6.4	6.2	11.7
44301	980	10.3	8.3	28.6
44515	2,415	6.5	6.2	8.1

Source: Bureau of the Census, 1990.

ing facilities provide jobs. What they may show is that the existence of a polluting facility may have an overall economic benefit for a region, but may not directly benefit the residents in the immediate vicinity of the facility. It may provide jobs for day workers who choose not to live there. Local residents may, in fact, hold few of the jobs, yet they and their families are exposed to much higher than average levels of toxic emissions. Without the trade-off of jobs for pollution, whether it be a health hazard or a nuisance, there can be little justification for the inequitable siting of polluting facilities.

State Policy and Environmental Justice

David Easton characterizes the political process as "a set of social interactions on the part of individuals and groups . . . (which) are predominantly oriented toward the authoritative allocation of values for a society."[13] Values are allocated in a society according to the power of the participants to gain advantage, as Easton points out:

> This is not to say that society as a whole need benefit from these settlements according to a given set of criteria of what is just or good. The order or regulation may, and typically does, favor one component group more than another.[14]

The poor and people of color are generally disadvantaged in the political process. As David Croteau explains, "Observers have long known that a correlation exists between high 'socio-economic status' (SES)—composed of education, income, occupation—and political participation in the United States."[15] Croteau argues that middle-class participants use a set of cultural tools that "are more useful and effective in navigating the political world as it is currently structured," and "working people often have access to a different set of cultural tools" that do not work as well.[16] As a result, the power balance in the political process is skewed toward those groups that have effective access to policymakers and bureaucrats and employ the "correct" cultural tools.

In the U.S. political system, values are generally allocated in accordance with socioeconomic status, as Croteau underlines. At the same time, the egalitarian philosophy that permeates U.S. politics requires decision makers to justify apparent inequities in the distribution of good and bad values. In the case of pollution, the justification is usually based on economic tradeoffs—the "victims" of the pollution also benefit from the jobs and economic development that the facility represents. While the entire community may gain from the polluting facility, the costs of the pollution usually fall more heavily on particular areas, which are almost always populated by the poor or people of color.

Ohio came of age during the industrial revolution, and, since the 1970s, many of its cities have not kept pace with the shift to a service economy. Consequently, the economic tradeoff argument remains powerful, so much so that it blinds policymakers to the possibility of solutions to problems like environmental injustice. The primary policy emphasis in the state is to protect indigenous industry and create new jobs. The environment is usually traded off in the absence of exceptional health effects data, and even that is not always persuasive.

A case in point is the trash-burning power plant located in zip code 43207 in Columbus. The first of its kind in the nation, the facility began operation in 1987 to burn a mixture of solid waste and coal in order to produce electricity. Several other sources of noxious fumes were already sited in the area, including the sewage treatment facility and a meat-rendering plant, since closed. The average income in the area is the highest in the sample at $27,389, but note that one-fourth of the residents are people of color and 42 percent of adults over eighteen have not completed high school. A significant proportion of the population are white Appalachians.[17]

In early 1994, William Sanjour, an engineer with the U.S. EPA, publicly revealed the results of 1992 stack tests at the incinerator, which showed dioxin emissions at levels more than ten times the allowable limit. These data had been suppressed by the EPA and the city of Columbus; when brought to light, they were swiftly denounced by both government and industry for scientific inaccuracy.

After pressure from citizens' groups and the media, a series of mass burns was conducted to more accurately measure typical dioxin levels. Even with evidence of the city's stockpiling of "safe" trash in an effort to skew the results, dioxin levels proved to be far above safe limits. The Franklin County Solid Waste District, which had taken over responsibility for the incinerator, appointed a committee to study the options and recommend solutions. The committee reported that the cost of retrofitting the facility to reduce dioxin to acceptable levels was too great. The trash burner was shutdown in December 1994, with a bonded indebtedness in excess of $175 million. The city health administrator still refused to accept the validity of the test data and the city began seeking a buyer for the plant.

The behavior of the state and city in this case illuminates the general perspective of policymakers toward polluting facilities in the state. The incinerator was sited in a neighborhood that already had a significant number of polluting facilities and low property values. It is also an area where residents have little political power because of income, education, and racial/cultural characteristics—a point of least political resistance. Very little attention has been paid over the years to the increase of toxic emissions in the area, whether because of official indifference to the residents or an implicit belief that pollution is the necessary tradeoff for economic development.

What is chilling is that even when presented with overwhelming evidence of toxic pollution far above acceptable levels, the state and city acted to protect the economic investment in the facility rather than to reduce the health effects on the population. The people in this neighborhood were clearly less important than the economic interests. There is no evidence that they were willing to make this tradeoff for the jobs themselves; indeed, they were not in possession of the necessary information to make such a rational choice.

The assumption that people in these areas willingly accept pollution has also served to limit the development of environmental policies in Ohio to those that have the least impact on industry. Policymakers often follow the path of least resistance: where powerful interests dominate the process, those interests take precedence. Focusing on business and industry interests rather than other values, such as human environments, results in disincentives to seek innovative solutions that are both equitable and cost-efficient.

One such solution, pollution prevention, is voluntary under Ohio law. In addition, when the state adopted legislation to implement Superfund Amendments and Reauthorization Act of 1986 Title III, the Community Right-to-Know Law, the threshold for reporting was set at 10,000 pounds for each chemical, matching the federal threshold. Lawmakers overlooked the experience of the Ohio cities that had already enacted right-to-know laws with lower thresholds. The city of Akron, for example, requires companies to report all chemicals that are stored, produced, or disposed of on-site in quantities of 500 pounds or more. The law was designed not only to provide citizens with information of toxic exposure but also to aid the fire department in case of an emergency at the location. The Akron fire department now has a database for each facility in the city that shows which chemicals are on-site and their precise location in the plant. This enables the department to have the proper equipment available to protect firefighters and area residents.[18]

An additional benefit of this lower threshold for reporting has been the incentive provided to local companies to reduce hazards. Being forced to reveal everything that they have on-site has increased their awareness of their liability and has led companies to rethink their production processes, inventory management, and disposal practices. Another incentive built into the law is a fee per pound assessed on companies. Fees in Akron pay for database management and monitoring by both the health and fire departments. Because the state's threshold is so high, the fees have not adequately funded implementation of the law in the counties. As a result, the quality of the information in most counties is inadequate.[19]

What this example suggests is that providing the right incentives can lead polluters to do what's best for society as well as their balance sheets. Requiring companies and subnational governments to focus on environmental justice as a goal may be the incentive for them to rethink the way they do business.

Pollution control has long focused on technology and economics, as if these were the only values that mattered. Yet environmental protection is aimed at reducing and preventing hazards to human health and the environment. When we consider trading off these values, we diminish their effectiveness as an incentive.

Conclusion

This study demonstrates that toxic polluting facilities in Ohio are overwhelmingly sited in areas where the residents are poorer, less educated, and have lower incomes and higher unemployment rates than the state norms. Whether this represents intentional racism or classism on the part of policymakers and industry decision makers is irrelevant. The inequity in siting is a reality—no toxic waste dumps are located in upper-income neighborhoods in Ohio or anywhere else. The fundamental issue is the base from which environmental policy decisions are made and the extent to which political inequities are perpetuated by them.

To address environmental injustice, policymakers must change their focus from economic values to the quality of life in all neighborhoods in the state. They must move away from blaming the victims for circumstances over which they have little control and adopt the premise that all people have the right to clean air, water, and a toxic-free environment, regardless of their political or economic power.

When that happens, environmental justice will become the fundamental policy goal; economic development will then be implemented in such a way that no neighborhood will be expected to absorb levels of pollution that are unwelcome in all neighborhoods. Industry will have the incentive to identify innovative technologies that prevent pollution, and government will design policies that reduce or eliminate the need for siting polluting facilities in anybody's neighborhood.

A new focus on environmental justice as a goal in and of itself may lead to the realization that economic and political rationales for continuing pollution in poor neighborhoods devalue all human beings. Forcing policymakers and industry to look at the social impacts of pollution may be the best way to identify policies and technologies that can enhance the quality of life and the environment in all neighborhoods.

Notes

1. Robert D. Bullard, "Solid Waste Sites and the Black Houston Community," *Sociological Inquiry* 53 (spring 1983): 273–88; Bullard, *Dumping in Dixie: Race, Class, and Environmental Quality* (Boulder, Colo.: Westview Press, 1990); U.S. General Accounting Office, *Siting of Hazardous Waste Landfills and Their Correlation with Racial and Economic Status of Surrounding Communities* (Washington, D.C.: General Accounting Office, 1983); Commission for Racial Justice, *Toxic Wastes and Race in the United States: A National Report on the Racial and Socioeconomic Characteristics of Communities with Hazardous Waste Sites* (New York: United Church of Christ, 1987).

2. U.S. Environmental Protection Agency Office of Environmental Justice, *Federal Register* 60, no. 51 (16 March 1995): 14283.

3. Christopher Boernor and Thomas Lambert, *Environmental Justice?* policy study no. 121 (St. Louis, Mo.: Center for the Study of American Business, Washington University, April 1994).

4. William M. Bowen, Mark J. Salling, Kingsley Haynes, and Ellen J. Cyran, "Toward Environmental Justice: Spatial Equity in Ohio and Cleveland," *Annals of the Association of American Geographers* (Albany, N.Y.: forthcoming).

5. The Toxic Release Inventory is a database compiled by the EPA on the location and amount of release of 320 toxic chemicals, as required by the Emergency Planning and Community Right-to-Know Act, Title III of the Superfund Amendment and Reauthorization Act of 1986. It is self-reported by the companies for levels of 10,000 pounds and above, and thus, underestimates total toxic releases in these areas.

6. See Bowen, "Toward Environmental Justice."

7. Boernor and Lambert, *Environmental Justice?*

8. William Ryan, *Blaming the Victim,* rev. ed. (New York: Vintage Books, 1976).

9. Data for the study were collected by intern Walter Heath at the Ohio Environmental Council in the summer of 1993 under the supervision of Richard C. Sahli, former executive director. The internship was supported by the Minority Environmental Summer Associate Program, the Environmental Careers Organization, and the Joyce Foundation.

10. The category *people of color* includes the following census designa-

tions: Blacks; American Indians, Eskimo, and Aleutian Islanders; Asian and Pacific Islanders; and other. In Ohio, 85 percent of people of color are Black.

11. In 1994, Ohio slipped to fourth, behind Tennessee.

12. The highest amount of emissions—77 million pounds—occurred in Lucas County, home to the city of Toledo. The county was not included in our study because none of its zip code areas contained at least 10 percent of county emissions.

13. David Easton, *A Framework for Political Analysis* (Englewood Cliffs, N.J.: Prentice-Hall, 1965), 50.

14. Ibid., 53.

15. David Croteau, *Politics and the Class Divide: Working People and the Middle-Class Divide* (Philadelphia: Temple University Press, 1995), 48.

16. Ibid., 47.

17. The census does not identify Appalachians as a class of whites. The neighborhood is known as the site of in-migration from the Appalachian areas of West Virginia and Kentucky. Because Appalachians are white, they are virtually invisible as a cultural entity in the census data.

18. See Mary Timney Bailey, Anthony P. Camma, and Thomas McDonald, "The Managerial Capability of Local Governments in Ohio to Address Hazardous Materials Problems," report for the Ohio Hazardous Substances Research, Education, and Management Institute, University of Cincinnati, Ohio, November 1992.

19. Ibid.

Linda Robyn and David E. Camacho

BISHIGENDAN AKII Respect the Earth

The process by which public policy is made is built on a variety of philosophical and epistemological arguments. An examination of these arguments is central to understanding the resultant policy. Since the policy approach is ultimately grounded in subjective choice, and is developed using political skills of strategy and persuasion, it is an intrinsically controversial activity. This chapter looks at how we think about the means and ends of policymaking. The central question is: What philosophical and epistemological frame of reference is best suited for policy leading to environmental justice? It discusses the limits of positivism and suggests that post-positivist inquiries are more appropriate for policy analysis. This proposition is tested by placing views about the environment into a Native American "way of life." In rethinking how environmental policy should be dealt with, this chapter stresses the significance of *values*. That is, there is a need to reconceptualize those values deemed to be "authoritative." Allocative decisions for society should be grounded in doctrines and principles that emphasize a collective, holistic way of viewing the environment. In short, we are in a state of environmental deterioration requiring alternative public policy approaches.

Positivism versus Postpositivism

Positivism is an epistemological school of thought that holds that scientific study proceeds from assumptions through deductive theory to the construction and empirical test of hypotheses, establishing verified causal explanations and general laws.[1] For much of this century, positivist philosophies of science demanded a reduction of all sciences, including social and behavioral, to the ontology and methods of physics. Much of social science adopted this thinking, following the belief that the questions and problems posed in the social world could be understood and solved using the same techniques as those applied to questions about the physical world. Once initiated into this rational, scientific method, the policy analyst (1) uses the theories, models, and concepts that structure scientific activity; (2) engages in causal explanations to illuminate the relations that hold between independent and dependent variables; (3) identifies patterns or irregularities in social and political life that can generate testable predictions; and (4) systematically tests such predictions against the contingencies of empirical events. Cost-benefit analysis, survey research, multiple regression, mathematical simulation models, experimental design, input-output studies, and systems analysis are techniques of the policy analyst schooled in positivism.

Some have come to question the ability of positivist approaches to deal with complex social issues like those considered in U.S. public policy.[2] The basic problem with the positivist approach is its inability to provide the analyst with a way to transcend political interest in order to obtain policy knowledge. That is, if the policy analyst cannot think beyond political interest in its various forms, then the policy thought process is dictated by that interest and is, therefore, fundamentally biased. In other words, public policy is dictated by ideological and financial considerations sympathetic to a dominant set of political interests. Deborah Stone observes that

> the rational ideal presupposes the existence of neutral facts—neutral in the sense that they only describe the world, but do not serve anybody's interest, promote any value judgments, or exert persuasive force beyond the weight of their correctness. Yet facts do not exist independent of interpretive lenses and they come clothed in words and numbers. Even the simple act of naming an object places it in a class and suggests that it is like some things and unlike others. Naming, like counting and rule making, is classification, and thus a political act.[3]

What is suggested, then, is how policy analysis might benefit from a methodology that acknowledges that scientific knowledge is dependent on the normative assumptions and social meanings of the world it explores. John Dryzek goes further in giving an active role to the policy analyst.[4] He suggests that the analyst not simply review and report on relevant factors, but also relate his or her own normative reflections to the study. Dryzek defines this "hermeneutic activity" as the evaluation of existing conditions and the exploration of alternatives to them, in terms of criteria derived from an understanding of possible better conditions, through an interchange between the "frames of reference" of analysts and actors. According to Dryzek, policy analysis should address ethics and normative theory *and* the apparent normative basis of the status quo in the decisional process—that is, the values and interest represented in the existing regime and policy process. In short, policy inquiry requires rigorous criticisms of policy proposals, *and* the implementation of policy must be defended before the public as the necessary and most important activity for producing the "public good."

In the same vein, Mary E. Hawkesworth argues that in order to effectively examine policy, the underlying values that drive decision making must be acknowledged.[5] Most significantly, for Hawkesworth, sources of power must be critically examined. Indeed, the critical study of any subject should take into account the hierarchies of power that are inherent in the society. This point speaks to the importance of *context*, whereby any subject would be better understood if it were examined within the setting that both produces it as a subject and brings it to light as something to be studied.

Our intent here is to challenge policy analysts to place themselves within a "frame of reference" grounded in the principles of environmental justice. The reader is directed to White (this volume) and the table therein identifying the principles of environmental justice. One frame of reference by itself does not inform the whole of the problems associated with negative environmental impacts on people of color and low-income groups. Moreover, we are asking the policy analyst to choose among social values, and because values underlie decisions, the policy analyst should recognize that the choice of a frame of reference is culturally bound and dependent—a point made by discussing the Native American "way of life."

A "Way of Life"

What alternative frame of reference is suitable for addressing environmental injustices? Described here is a new frame of reference grounded in the doctrines and principles of Native American views on the environment. The Native American perspective demands critical thinking about the policies of the private and public sectors developed in response to environmental issues. It questions the assumptions on which those policies are based, examines traditional solutions, and advocates new ways of thinking about the environment. Questions are asked about the responsibilities toward the environment, and how these responsibilities ought to be reflected in the policies adopted by the government and private companies as well as in the habits of the population as a whole. The idea of this section is to inquire and advocate; to critique the positivist foundations of the public policy process.

Respect the Earth

For the Ojibway people, the environment is not an issue. It is a way of life. As with other tribes, the Ojibway consider themselves inseparable from the natural elements of their land. Environmental sustainability is the "ability of a community to utilize its natural, human and technological resources to ensure that all members of present and future generations can attain a high degree of health and well-being, economic security and a say in shaping their future while maintaining the integrity of the ecological systems upon which all life and production depends. The four pillars of sustainability are: economic security, ecological integrity, democracy, and community."[6]

This definition of environmental sustainability is relatively new, but many Native Americans have been practicing the "new" concept for a very long time. An example is Archie Mosay, a ninety-year-old Ojibway elder, traditionalist, and spiritual adviser from the St. Croix Reservation in Wisconsin who teaches by word and example. His lessons on the environment come from the traditional values taught by his parents and grandparents. Mosay's teachings pass down his belief that tribal members are responsible for working to protect the earth for the generations that will follow. Speaking from his rich traditional background, Mosay gives this message to those who are concerned about what can be done to preserve native environments and values:

I don't know where exactly I was born. I lived in a wigwam until I was eight or nine. I've hunted since I was 12. Our chief said not to take any more from the earth than we would use. And there wasn't much waste. Indian people didn't kill for sport—it was to live and to survive.

Try to live with respect for the Earth. Live the way I did when I was young. There was community. Elders talked to the young people about how they should live their life.

Whatever grows on this earth belongs to the Indian people. The spirits put that there for the Indian people to use. Take care of the environment the way my grandparents taught me. Go to your elders and listen. Pray for the Indian people.

There are a lot of things the younger generation forgets about. They've got too many things in their minds. Don't forget about Indian culture. Don't forget about fasting, about sweat lodges.

Pray. Live right; and heal the earth.[7]

Looking even further than Mosay, Mishi-Waub-Kaikaik gives us an ancient insight into Ojibway values that influence our current views of the environment with this appeal on consideration of a proposed treaty:

Can man possess a gust of the North Wind or measure of flowing water? Can he control a mass of clouds or a herd of moose?

No. Do not mistake the truth. It is not man who owns the land; it is the land that owns man. And we, the Anishnabe, were placed on this land. From beginning to end it nourishes us: it quenches our thirst, it shelters us, and we follow the order of its seasons. It gives us freedom to come and go according to its nature and its extent—great freedom when the extent is large, less freedom when it is small. And when we die we are buried within the land that outlives us all. We belong to the land by birth, by need, and by affection. And no man may presume to own the land. Only the tribe can do that.

Do we have the right to sell our homeland? Do we have the right to rob our children of their claim to it? Do we have the right to sell the resting places of our ancestors? Do we have the right—even if we agreed—to sell what is a gift of the Great Mystery, and what also belongs to those who will come after us? I cannot think so.

For generations our forefathers freely ranged this land, lighting their fires where they wished and burying their dead in places now sacred. During our lives we, too, have enjoyed the land and its freedoms and our children and their children should have the same rights. What I wish for you and for our children is a place where we can abide in peace and without want, where we can watch our children grow and hear them laugh, where we can smoke the pipe of friendship in our old age, and where we can die together and be buried near our ancestors.[8]

Native American holistic views of the environment came into conflict with the dominant capitalistic nature of early European settlers and continues to do so today. Since the beginning of the U.S. republic, control of the land and natural resources have been a constant source of conflict between Euro-American settlers and indigenous nations. Disputes over land usage and ownership have defined the totality of government–Native American relationships from the first contact to the present day. The white person's perspective of the exploitation of land and its resources will continue into the foreseeable future. Mining projects, development proposals, and get-rich-quick schemes have been inflicted on tribes for years. Millions of dollars are at stake, with large multinational corporations and the federal government clamoring to do business on reservations.

The 287 Native American reservations within the United States, from Florida to Wisconsin to Alaska, are among the most exploited and environmentally degraded lands anywhere in rural America. With the sanctioning of certain power arrangements by the federal Bureau of Indian Affairs, corporations and federal agencies have pressured, bribed, cajoled, and enticed their way in to mine for strategic minerals that would environmentally devastate the sacred rice beds of the Sokaogon Chippewa; to strip-mine coal, as on the Crow and Navajo reservations; to drill for oil, as on the Blackfeet reservation; and to site garbage dumps and medical waste incinerators, as on the Salt River and Gila River reservations. This process of exploitation and expropriation goes on and on.[9]

The belief that natural law is supreme, and should provide the guiding principles on which societies and peoples function, is what distinguishes the Native American perspective on the environment from the dominant paradigm of Eurocentric environmental exploitation. The holistic view of sustainability for the Ojibway people is that laws made by nations, states, and municipalities are inferior to natural law, and should be treated in this manner. Many Anishnabe people abide by a code of ethics and a value system that harmonize human behavior with natural law. Translated as the "good life" or "continuous rebirth," *mino bimaatisiiwin* guides behavior toward other human beings, animals, plants, and the ecosystem, and has as its basis the tenets of reciprocity and cyclical thinking.

In contrast to the dominant paradigm, reciprocity is a responsibility, as well as a way of relating between humans and the ecosystem. Because the resources of the ecosystem (plants, animals, water, and air) are viewed as animate gifts from the Creator, a person cannot take one of these gifts without a reciprocal offering, usually tobacco or *saymah,* as it is called in the Ojibway language. Within this understanding of natural law is the belief that one takes only what is needed and the rest is left. Under natural law, most of what is natural is cyclical, as is evidenced by our bodies, the seasons, and life itself. This natural cycle clearly gives us a sense of birth and rebirth, and the knowledge that what one does today will affect the future as the cycle completes itself.

The basic practice of *mino bimaatisiiwin* helps us understand the relationship between humans and the ecosystem, and our need to maintain balance. Social and economic systems grounded in these values tend to be decentralized, communal, and democratic, allowing the people to rely on the land of that ecosystem.

Holistic environmental paradigms stand in sharp contrast to life in an industrial society. Natural law is preempted as "man's domination over nature" becomes the central way of life. Rather than the Native American cyclical process of thinking, a linear concept of progress dominates industrial societies. Progress is defined in terms of economic growth and technological advancement, and is key to the development of dominant "civilized" societies. From this perspective, the natural world is seen as something that is wild and in need of taming and cultivation. Those not part of this mentality were, and are, seen as primitive and in need of being civilized. "Civilizing" those not part of the dominant paradigm is the philosophical basis of colonialism and conquest.

Social interpretations of this linear, scientific way of thinking, such as Darwinism and Manifest Destiny, have resulted in particular actions in which some humans believe they have the God-ordained right and duty to dominate the earth and other peoples. As Winona LaDuke argues, the difference in these two paradigms demonstrates the scope of the problem and the reality that a society based on conquest cannot survive.[10]

LaDuke reports that during the last 100 years, industrial society has caused the extinction of more species than in the period between the Ice Age and the nineteenth century, and in the past 400 years, an estimated 2,000 indigenous peoples have been made extinct. Moreover, she contends the native peoples, both worldwide and in North America, are at the center of the present environmental and economic crises, and not coincidentally. Even though native peoples represent a demographic minority of the total population in North America, for instance, they still maintain land occupancy over substantial areas of the continent. LaDuke points out that in many regions of the United States, native peoples are in the majority, such as in parts of New Mexico, Arizona, northern Minnesota, the Dakotas, and Montana. The native population also forms the majority in two-thirds of Canada, a landmass roughly one-third of the North American continent. Much of the northern population is native in areas such as Quebec, Newfoundland, Labrador, Ontario, and the west coast in British Columbia. The native population of the Arctic and subarctic is substantial as well.

In this context, Native American perspectives regarding native thinking, survival of native communities, issues of sovereignty, and control over natural resources are becoming central to North American resource politics.

LaDuke, furthermore, offers the following insights:

- Over fifty million indigenous peoples inhabit the world's remaining rainforests.
- Over one million indigenous peoples are slated to be relocated for hydroelectric dam projects in the next decade.
- All nuclear weapons that have been "tested" by the United States have been detonated in the lands of indigenous peoples, with over 600 tests within the Shoshone nation alone.
- Two-thirds of all uranium resources within the "borders of the United States" lie under native reservations, with Native Americans producing 100 percent of all federally controlled uranium in 1975.

- Fifteen of the present eighteen so-called monitored retrievable nuclear storage sites are in Indian communities.
- The single largest hydroelectric project on the continent—James Bay—is on Cree and Inuit lands in northern Canada.

There are many issues in the web of struggle between industrial and indigenous societies, and as LaDuke points out, it is a struggle that ultimately encompasses us all.

Even though Native American perspectives are beginning to inform environmental politics and policy to a greater extent, at present, Native American philosophies and values are not included in those policy decisions that benefit large corporations and serve the interests of the state. There is a vast expanse of social distance between all involved that causes a breakdown in communication as well as misinterpretation of each other's actions. This social distance is well illustrated by Walter Bresette, activist and member of the Red Cliff band of Chippewa, who says that Native Americans and non–Native Americans alike are being victimized by large corporations, which reduce economic options.[11]

As activist and author Al Gedicks writes, "The sooner we stop labeling 'native issues' as something separate and distinct from our own survival, the sooner we will appreciate the critical interconnections of the world's ecosystems and social systems."[12] Environmental concerns can be absolutely crucial within the context of reservation politics; even before the most hostile of tribal councils, the kind of "Mother Earth" talk that would make Anglo mining executives or legislators roll their eyes can make all the difference.[13] Corporate America and the federal government would be wise to realize that there is a growing respect for tribal elders and the "old ways." Utilitarian business practices and government actions that benefit all involved cannot be accomplished by ignoring this fact.

Mining and Treaty Rights Conference[14]

How, then, is it possible to avert exploitation, misconceptions, and future environmental disasters? Bresette argues that "the answer lies in balance—the same today as was taught centuries ago." Speaking at the 1992 Mining and Treaty Rights Conference in Wisconsin,

Bresette stated that "the reason the mines [here] are functioning has less to do with the economy and more to do with the imbalance that exists. Long term strategy, looking forward to the seventh generation, has much to do with balance; how it affects our psyches as well as legislation."

Moreover, Bresette continued, "when balance is achieved, effective strategies will come about. Take what is out of balance, such as anger and rage, turn it around, and use it strategically. Balance is necessary to sort out bad and evil as well as to realize our limits."

In speaking about the earth, Bresette said that "land is the issue—reclaim the land. Earth will be real happy when we're all gone. If we don't change, things will get worse." Again, this refers to the social distance and difference in values between many Native Americans and mainstream American culture. One does not own the earth and the ecosystem that it is a part of; rather, as was learned generations ago, humans should exist in harmony with the earth. This concept needs to be revived in order to ensure the health and balance of children yet unborn.

Tom Maulson, tribal chair of the Lac du Flambeau Chippewa, opened the mining conference by stating, "Another day we have to fight for everything. Should we as Anishnabe people help Whites in this struggle? We have to help because mining corporations are everywhere. We have to take on the task of leadership and support people. Indians are a true minority, but have treaty rights which we have to protect come hell or high water. All corporations are here to mine, and we will stand together to get them out. By doing what they're doing today to Mother Earth, there won't be a tomorrow. They won't have to worry about Indian spears going into the water for fish because there won't be any."

During the conference, Fred Ackley, tribal judge of the Mole Lake Sokaogon Chippewa, explained what happened when his people learned that they could make a lot of money by allowing Exxon to mine on their reservation. As a tribe, they assessed the experiences of other tribes that had leased their lands and were short-changed. After weighing the consequences, the Sokaogon decided not to pursue mining as an economic possibility on their tiny, 1,800-acre reservation in Wisconsin.

Ackley said, "We look at Exxon as a battle, but we are getting tired, worn out of fighting 200 years of U.S. promises. Our grandfathers fought alongside the French and British, and now I need to fight for my grandsons and teach them not to develop [the land] so the water can be purified." He used the Indians along the Sturgeon River mine in Wisconsin as an example. "They didn't consider the pollution of the mines and now the water is forever contaminated. They shipped copper from Keeweenaw and sent it to France; it came back as the Statue of Liberty. . . . The ceded land is now owned by Exxon and other large corporations. Much mental anguish comes from corporations when they make people think they can't fight."

When the Sokaogon said "no" to mining, Ackley pointed out, they said "no" to wealth.

Ackley's tribe "looks at uranium as a monster that's getting us all. There is no word for minerals in Chippewa, no specific translation, so how can they say the Chippewa leased this land? We are supposed to stand by and let corporations take what they want, and because we fight we are called activists." Ackley and other members of his tribe are willing to use nonviolent but unconventional methods of resistance against the state and Exxon even if there is a high possibility of going to jail. "We've been fighting a long time and have to fight a long time more because corporations own mineral rights," he added.

For Ackley, however, the money that could be gained from letting Exxon and other corporations mine means nothing if the pollution that results causes people to contract diseases such as cancer. Money also means nothing when faced with the possible loss, if Exxon is allowed to mine, of the only remaining lake with ancient rice. Rather than seeing humans as being one with the earth, Ackley observes that corporations such as Exxon think and act as if they are not a part of the earth. "There are big spirits in the copper trying to warn them." Lastly, he is afraid that if the Sokaogon fight for natural resources, even that will be taken away. The treaty forming their reservation was drawn up in the 1930s. If the Sokaogon do not yield to the corporations, then according to this treaty, the federal government through its use of plenary powers—the doctrine that views Native American country as "owned" by the federal government—can force them to leave their reservation.

Mike Sturdevant, leader of the Menominee Warrior Society, states that "if it were just the Whites, they would not be able to stop the corporations. Whites need Indians to stop mining because Indians have treaty laws, and treaties protect the land." In the traditional warrior spirit, Sturdevant believes that his tribe must fight to protect what little is left and expressed his willingness to intervene to stop the corporations, even if it costs him a prison sentence. Even passive resistance is met by the threat of or actually going to jail.

In the tradition of the old ways, Sturdevant recounted a vision that came to him during a ceremony. In this vision, a spirit told him that the government was trying to take the last of the Menominee's resources—land, water, minerals, and timber—by bribing tribal and government officials. If bribery fails, and as a last resort, there is a congressional document in Washington, D.C., according to his vision, that details how to "kill the rest of the Indians off."

Policy Alternatives

To corporate executives and government legislators, the vision recounted by Sturdevant is just that: a vision. The point is that corporate executives and elected officials do not understand the deep, spiritual relationship of the Chippewa to their land. Instead, the corporate-government misconception is that everyone would like to be wealthy. Existing on $400 a month, when one could live quite comfortably on corporate dollars, makes no sense to those who sit in corporate boardrooms. If corporate and government leaders do not understand by now, they probably never will. But the message is clear. If they could become more accepting of Native American beliefs, perhaps they could learn that when one is involved with Native American people, there is a spiritual side and an interconnectedness with the earth that cannot be ignored.

Environmentally sustainable development is inseparable from maintaining cultural diversity. Reopening and broadening the public debate about the economic and environmental future of indigenous peoples would allow input from groups that are normally ignored in the decisional process. Allowing for consideration of alternative plans would offer a different means of economic development in places like northern Wisconsin while challenging the traditional export-based models of economic development, because mining and oil and gas drilling are extremely capital intensive. That is, if the goal is to provide jobs and a stable rural economic environment, investing in mining, oil, and gas is exactly the wrong way to accomplish this task. Mining industries that provide jobs for only a short period of time and that also pollute are a poor investment in the long term. Small, locally owned firms and labor intensive ventures—such as tribal fish hatcheries, renewable energy, recycling, forest products, and organic farming—would create far more jobs than mining while contributing to an environmentally sustainable economy.[15] For example, Menominee Tribal Enterprises in Keshena, Wisconsin, received international recognition for its achievements in sustainable forestry. The Menominee manage 220,000 acres of forested lands and are now an acknowledged "leader in shelterwood systems for uneven-aged management of white pine, hemlock, and hemlock-yellow birch ecosystems."[16]

Cost-effective technologies for utility companies to buy and use locally owned renewable energy sources are now available. Rather than encouraging electric companies to build six proposed coal-fired power plants, as happened recently, the state of Wisconsin could have encouraged the use of these much less polluting technologies. Not only would the economy benefit, but less mercury, which enters fish through acid rain, would be released into the atmosphere.[17]

An alternative to these short-term, wasteful corporate and government projects would be a type of sustainable development that recognizes "the right of each indigenous people to a land and resource base necessary to sustain an appropriate and sufficient economy and the right to exercise its authority and jurisdiction over the corresponding territory."[18]

Conclusion

Industrial societies generate a much greater abundance of material items than traditional societies. One could, however, make the argument that a sense of personal fulfillment, control over one's time, and a general peace of mind—a quality of life realized within traditional native societies—greatly outstrip the material goods of industrialized societies. Viewed from this perspective, it would seem that the quality of life for many deteriorates as the degree of industrialization progresses.[19] Providing for substantive public participation in policymaking decisions, and allowing other views and values into this process, would help even the balance. Unfortunately, for the most part, projects are planned out and presented in such a way as to make the public feel that their input is unimportant and would not make a difference anyway.

With the assertion of treaty rights and traditional Native American values, resistance to ecologically destructive projects is being taken more seriously. Rather than focusing solely on how projects will be developed, the emphasis is shifting to include who will be involved in the decision-making process on a more frequent basis. Corporate-government alliances have been slowed down as indigenous peoples and non-natives have voiced concerns about the social, economic, and environmental impacts of these projects.

Both environmental protection and devastation follow the path of least resistance. Environmental harms are connected to many issues, such as the air we breathe, our food, water, lifestyles, and legal decisions. Developing environmentally sound, long-term, sustainable economic alternatives will depend on many variables, such as research, organizing and lobbying, legal representation, effective use of the media, interactive skills involving native rights and environmental movements, and an earnest inclusion of native beliefs and values—singularly or in combination—in the political processes of deliberation and allocation. With the inclusion of these values, socially harmful interactions between economic and political institutions can be decreased while restoring the balance that is so important to native peoples. Clearly, this set of values challenges the harmful and wasteful projects of profit-maximizing corporations and growth-at-all-costs government policies.

These values are not incompatible with those of the growing movement around issues of environmental justice, especially in relation to the quality of life. It is crucial that Eurocentrism be reassessed for its impacts on the environment, tradition, and native peoples. Considering Native American views and philosophies concerning the environment can help accomplish this by creating a new heritage of respect, cooperation, and freedom. This is not an easy task, but we must begin to take a more holistic approach, and learn to mimic and live in harmony with natural systems. We must all learn to have respect for and afford justice to other humans as well as other living species. We must learn that we are all related; that is, that all things are related.

Notes

1. Deborah Stone, *Policy Paradox and Political Reason* (Glenview, Ill.: Scott, Foresman, and Company, 1988).

2. Frank Fisher and John Forester, eds., *The Argumentative Turn in Policy Analysis and Planning* (Durham, N.C.: Duke University Press, 1993).

3. Stone, *Policy Paradox*, 252–53.

4. John Dryzek, "From Sciences to Argument," in Fischer and Forester, *The Argumentative Turn in Policy Analysis and Planning*, 214–15.

5. Mary E. Hawkesworth, "Epistemology and Policy Analysis," in *Advances in Policy Studies since 1950*, ed. William Dunn and Rita Mae Kelly, vol. 10 of *Policy Studies Review Annual* (New Brunswick, N.J.: Transaction Publishers, 1992), 295–329.

6. Anthony Cortese, E. Kline, and J. Smith, *Second Nature Partnership Training Manual* (February 1994) (East Cambridge, Mass.: Second Nature).

7. Published by the Great Lakes Indian Fish and Wildlife Commission, P.O. Box 9, Odanah, Wis. 54861 (no volume number or date of publication given). The commission is dedicated to the protection and preservation of Chippewa treaty rights and natural resources.

8. Basil Johnston, "Ojibway Ceremonies," in *Seasons of the Chippewa* (Odanah, Wis.: Great Lakes Indian Fish and Wildlife Commission, 1993).

9. Margaret Knox, "Their Mother's Keepers," *Sierra* 78, no. 2 (March/April 1993): 50.

10. Winona LaDuke, "A Society Based on Conquest Cannot Be Sustained," foreword to *The New Resource Wars*, Al Gedicks (Boston: South End Press, 1993).

11. Walter Bresette, notes from the Mining and Treaty Rights Conference, Tomahawk, Wis., October 1992.

12. Al Gedicks, *The New Resource Wars* (Boston: South End Press, 1993), 202.

13. Knox, "Their Mother's Keepers," 50.

14. The following comments from Walter Bresette, Tom Maulson, Fred Ackley, and Mike Sturdevant were made at the Mining and Treaty Rights Conference, Tomahawk, Wis., October 1992. Linda Robyn has

quoted them as accurately as possible from her notes and the public record.

15. Gedicks, *New Resource Wars,* 197.

16. Ibid.

17. Ibid.

18. Ibid., 200.

19. M. Annette Jaimes, "Re-visioning Native America: An Indigenist View of Primitivism and Industrialism," *Social Justice: A Journal of Crime, Conflict, and World Order* 19, no. 2 (summer 1992): 20–21.

David E. Camacho

ENVIRONMENTAL ETHICS AS A POLITICAL CHOICE

This chapter attempts to synthesize some of the discussion in the preceding chapters. And like the previous two chapters (Timney; Robyn and Camacho), another frame of reference is introduced for students of politics and public policy to ponder. It should be noted here that one frame of reference by itself does not inform the whole of the problems associated with negative environmental impacts on people of color and low-income groups. Instead, the *context* under study very much determines which frame of reference may be most suited for addressing the particular problems. Contrast Timney's discussion of the incentives required for influencing the "economic" behavior and practices of the business community with Robyn's insight into the Native American "life view" of the environment. Timney is calling for changes in individual behavior and attitudes, whereas Robyn is addressing a cultural, collective way of viewing the environment. Both stress the significance of *values* in rethinking how environmental injustices should be dealt with.

This chapter also stresses the need to reconceptualize those values deemed to be "authoritative." Allocative decisions for society must be grounded in the doctrines and principles of environmental ethics. Rogene A. Buchholz suggests that as a form of applied normative ethics, environmental ethics deals with the approach that *ought* to be taken (Robyn and Camacho) as well as the approach that *is* taken (Timney).[1] Principles of environmental justice, in turn, must be adjoined to those of environmental ethics. The reader is directed to White (this volume) and especially the important table identifying the principles of environmental justice, which can be defined as "environmental protection aimed at equalizing social opportunity."[2] Furthermore, "environmental justice is the belief that both environmental benefits and environmental costs should be equally distributed in society and that corporations should be obligated to obey existing laws the same as individuals are so obligated."[3] Through ethical behavior, justice will be served.

Background

As noted in this anthology, individuals and groups have promoted protection of the environment since at least the 1830s. Today, the threat to human survival is of a global dimension. Although many individuals and groups have been attracted to the modern environmental movement, mainstream environmental groups have been slow in broadening their base to include people of color, the poor, and working-class whites. There are efforts at coalition-building, but conflicts remain over how to balance economic development, social justice, and environmental protection. Hindering efforts at coalition building is the lack of concern over "justice" in the mainstream environmental movement. It is true that prior to the 1980s, there was little involvement by people of color in environmental activism for two main reasons. First, the mainstream environmental movement was overwhelmingly composed of white, middle- and upper-class members whose principal focus was preserving natural areas and endangered species. Second, people of color simply did not have the time to give primary concern to their environment when confronted by pressing problems related to their day-to-day survival. Poverty, unemployment, poor health, and inadequate housing overwhelmed concern for the environment.

By articulating its view of environmental problems in terms of racial discrimination and social justice, the environmental justice movement has built on the civil rights movement. Much of the success of the environmental justice movement in the United States can be attributed to the ability of the movement to align itself to the legacy of the civil rights movement. This proposition is supported by social movement theory and the political process model, which identifies factors for successful movement insurgency. In the end, environmental justice advocates are building a multiracial and inclusive movement that has the potential of transforming the political landscape of this nation. The movement is inclusive in that environmental concerns are not treated as separate from health, employment, housing, and educational concerns.

But sustaining a coalition based on racial, gender, and class differences is a formidable task. Limited resources impede organizing; lack of information can block mobilization efforts; determining accountability and responsibility for the environmental hazard can be impossible—these are some of the difficulties facing the movement.

What is to be done? What public policy choices adequately address environmental injustices? What authoritative values should govern the decisional process? This chapter discusses a "frame of reference" that may lead to effective environmental planning and policy. Citizen deliberation is essential if there is to be a collective effort addressing environmental injustices. Because the environmental crisis is a political crisis, thinking about the environment must be cast in political terms. A change in individual attitudes and behavior is necessary before deliberations begin or attempts at coalition building are made.

Environmental Ethics and Justice

Environmental activists and theorists have been slow to develop a theory of political action because of their focus on instrumentalities: rules, bureaucracy, and administration in general. What is needed is a cooperative effort between corporations, governmental bodies, and grassroots organizations. In the end, individual behavior, values, and attitudes must be reassessed according to doctrines and principles grounded in environmental ethics and justice. The central question here is: What is to be done?

As a starting point, the cultural, political, economic, and social factors that shape human behavior in society must be understood. The view taken here is that environmental problems are a result of determined, regularized structures of society. More precisely, the values of the "consumer society" in which we live have played a major role in creating a wide range of environmental problems. While our focus is on U.S. society, these cultural values predate this country and have long been characteristic of the Western world.[4] This dominant Western worldview can be understood in terms of people feeling separate from and superior to nature (Robyn's principal thesis). It is not only an antiecological viewpoint; it extends into the realm of ethnocentrism, with its corresponding propensity to deprecate the ways of persons from other societies as wrong, old-fashioned, or immoral, and to value the ways of one's own group as superior (a major thesis in Berry's discussion of Eurocentrism, this volume). When the belief that your group's way is the only right way of doing things enters the authoritative and legitimate structures of society, then practices leading to discrimination by one group over another follow. Berry traces this phenomenon in her assessment of legal doctrine. Indeed, the unfavorable treatment of individuals because of their group membership has been the central focus of *Environmental Injustices, Political Struggles.*

U.S. Culture: A "Consumer Society"

A brief overview of some of the specific dimensions of U.S. culture that contribute to an antiecological perspective is in order.[5] First, rather than living in harmony with nature or submitting to nature, we in the United States have placed the most value on *overcoming* or *dominating* nature. We take great pride in the "taming" of the New World. This "taming" not only transformed the land but Christianized and "civilized" the native "savages" as well. The practice of genocide and the removal of Native Americans from their homelands have consequences for contemporary environmental politics. For example, Berry provides historical evidence to show how the abuse of the human condition continues. The historical variable is given further significance by Bath, Tanski, and Villarreal (also this volume) in their assessment of the *colonias* of El Paso County. Present practices reflect the past.

Second of all, we place a high value on *progress,* that is, we in the United States believe there should be progressive, steady improvement throughout our lives. We tend to be future oriented to the extent that the future should hold further opportunities for progress. Progress usually means greater technological success in creating machines, buildings, communications systems, and so forth.

Lastly, *materialism* is an important cultural value in the United States. People want more and more things. A critical distinction must be made between "wants" and "needs." Individuals do not need more "stuff," they simply have an insatiable appetite to consume, often to impress others. This childlike behavior to stay ahead of the Joneses has drastic consequences, primarily because of the belief that we have a divine right or entitlement to take from the environment because of our ability to dominate it.

These values combine to lead individuals to exploit and ultimately abuse their environment with little or no regard for the negative ecological effects. The fact is that as a culture, we regard control over nature, progress, and material goods as far more important than the ideal of living in harmony with nature. Thus, a significant factor contributing to our environmental crisis is our value system, and major changes must take place to reprioritize that value system.

Until such a reordering occurs, we are likely to do little more than respond to each crisis as it arises. But a piecemeal approach to solving complex problems will not suffice. Recently, we have attempted to come to terms with environmental problems, but only because we have been forced to do so. That is, the environmental problems that we have created have begun to threaten our health and even our lives. Out of our own self-interest, then, we are placing more value on a healthy and thriving environment. As Robyn and Camacho suggest, "quality of life" issues may bring together seemingly disparate perspectives on the environment.

Attention is now given to how one's self-interest, or individual behavior, can—in small, cumulative ways—alter the pressure society places on ecological systems. Self-interest, the reader is reminded, is a strong motivator for the individual to take political action. And in the end, political action is what will be required to create a balance between extraction and preservation; political action, in other words, will determine our level of environmental sustainability. Perhaps we can accomplish such a task if the impetus for this balancing act is the "quality of life."

The Individual as Rational Actor

The preponderance of U.S. culture to exploit and destroy the environment speaks to how social structures influence the individual's thinking and behavior.[6] This structuralist view is not so deterministic that change cannot occur. Changes in one's behavior, both in itself and, more important, in conjunction with similar changes in others' activities, can stimulate societal changes. The aggregation of individual behaviors can alter the structure of society. Policy models, then, must accept and understand the major dimensions of U.S. culture that currently produce many environmental problems. For example, changes in the routine economic operations of society must be assessed for their impact on the environment.

Considered below are practical, commonsensical ways that individuals can change their attitudes and behavior toward the treatment of the environment. The individual is assumed to be a rational actor, acting out of his or her own self-interest. This is not an argument for a pure economic application of a cost-benefit analysis.[7] Rather, it suggests that it is theoretically possible that a cost-benefit analysis undertaken by an individual will result in that individual working in a collective manner toward the betterment of the environment as a way to protect self-interest.[8] In turn, understanding the collective aspect of the "quality of life" issue can result in group action that challenges the human dimension discussed in this volume. That is, the human condition *(race)* and the ecological crisis *(environment)* cannot be treated as separate problems. The Native American, indigenous belief that all things are related must *frame* the comprehensive, holistic planning around the complexities of the environmental crisis.

What practical proposals can be offered to the consumer?[9] The individual can be more ecologically responsible by buying fewer goods, purchasing goods that involve more recycled materials and are themselves recyclable, using more durable or reusable goods, purchasing used items wherever possible, and passing on items that are no longer needed to social service organizations for recirculation to others. As a citizen-consumer, one can organize to help the community to reuse and recycle more effectively, perhaps by concentrating some remanufacturing activities at the community level to create employment. One can also use the vote to support candidates and organizations that promote more social control of reusability, recyclability, incorporation of recycled goods into products, and pollution control at remanufacturing facilities.

What implications do these guidelines pose for public policymakers? This question speaks to the crux of this chapter, for economic factors must be taken into account by policymakers. As suggested by Timney, the economic policies that must be considered may require a variety of incentives (subsidies) and disincentives (taxes and fines) from governments to redirect production organizations to act in more ecologically responsible ways.

What actions might government take? Clearly, governments need to regulate the economic marketplace. Several responses are available: (1) direct control of access to ecosystems; and (2) economic influence over markets through (a) the levying of taxes, (b) the imposition of fines, (c) subsidies for specified forms of ecosystem usage, and (d) various price controls over market producers. The market will not routinely solve environmental problems given that the primary concern of business is economic performance. Business cannot set aside the "bottom line," or its profit motive, to pursue objectives that may conflict with economic performance; it will not survive the competitive nature of the marketplace. Without government intervention, economic elites are not likely to routinely apply an ongoing and long-term ecological outlook to their decisional process. Government should also provide the financial and technical support for ecological monitoring research because of the prohibitive cost to private institutions or individuals. Research findings must then be disseminated by government to a variety of citizens movements, political organizations, and economic decision makers. These ideas do not constitute the whole of governmental regulations and activities, but rather, should be the minimum tasks of government.

The pluralist perspective on power (Camacho, beginning of this volume), which helps explain the conditions for effective political pressure, suggests that governments respond to actions taken by well-organized, politically knowledgeable, and effectively mobilized groups. But every government can take a variety of stances toward political pressure. As regards responses to environmental activists, some governments will reinforce the authoritative cultural values of progress, materialism, and domination over nature held by economic elites, and at the same time, will discourage the protests of most environmentalists to modify the market systems of the economy. Other governmental units will be formally neutral, neither discouraging nor encouraging government regulation of the economy. Finally, some governments will be encouraging of many environmental protests, siding with such groups over the resistance of many economic elites. Among the factors that explain how governments will respond are (1) the state of the economy, especially its level of surplus and potential for growth; (2) the strength of the environmental activist movement; (3) the degree to which other protest movements are in congruence with or in opposition to environmental protests—for example, social welfare or poor peoples' movements; and (4) the nature of scientific and popular understandings of environmental hazards.[10]

There are also more expressly political behaviors, emphasizing the individual's role as citizen rather than economic consumer. How should the individual as an environmentally concerned citizen respond to the problems discussed here? There is ample information in *Environmental Injustices, Political Struggles* to draw from in deciding how to respond as a citizen activist. For example, understanding the significance of the political variable is imperative.

The framework or model that stems from this most important of variables explains environmental problems as a result of many institutions and powerful actors having a strong self-interest in extracting "resources," limited as they may be, from the environment. Because of different cultural perspectives and life views, it is not practical to envision a situation wherein we relate to the environment in ideal harmony. Instead, there will be inherent conflicts over how we want this relationship to be structured. It was suggested above that cultural values were at the root of these conflicts, and that this clash of values would be reflected in political, economic, and social disputes. Governmental regulation of the marketplace will restrict access to the environment, thus exasperating the condition of "limited resources," or scarcity. The basic issue to resolve, therefore, is who will effectively pay for this scarcity of access—through taxes, fines, jobs, wages, social welfare reductions, or health risks. *Environmental Injustices, Political Struggles* has pointed out that, heretofore, people of color and the working poor have carried a disproportionate amount of the cost of public policy decisions.

In short, the citizen activist must understand factors that have contributed to the environmental crisis. Conflict surrounds environmental issues. Accordingly, the political variable must be studied carefully, for in the end, political solutions must be found to deal with the environmental crisis.

It should be noted that it is possible to participate in environmental activities that are nonconflictual, such as developing new parks, planting trees, and recycling. These activities, however, do not achieve much in the way of environmental protection. Moreover, it is safe to predict that almost every grassroots environmental movement will confront well-organized and persistent opposition from a variety of economic and political actors. This resistance may take the form of threats or "symbolic politics" on the part of economic elites—for example, the loss of jobs or green marketing by producers. Grassroots environmental movements may be confronted by groups of workers or the poor (whose positions and protestations may be supported by economic elites) because of the threat to their financial well-being.

What practical proposals can be offered to the citizen activist?[11] First, if one cannot function well under conflict, be accepting of other environmentalists who can and do function within an inherently conflictual political system. Second, if one cannot or does not want to participate in such political conflicts, the individual can still offer political support for candidates and movements that have serious proposals for reforms leading to environmental protection—for example, by voting or through financial contributions to such movement organizations. Third, the citizen activist can be alert to issues of environmental justice. What are some ways, for instance, to distribute the burdens of environmental protection more humanely? Be aware of these ways in family settings, schools, workplaces, or communities. It is crucial that the "human dimension" be addressed.

Even with these practical proposals, there is no easy answer for "what is to be done." In the modern, capitalistic society of the United States, all individuals require the extraction of some resources from the environment for their subsistence. But left unchecked, economic pressures will overwhelm most ecological systems. This is especially true in the modern, industrial (and postindustrial) era, when successful extraction leads to increased technological and economic power to accelerate the "treadmill of production."[12] Under this pressure, the environmental problems identified in *Environmental Injustices, Political Struggles* will only increase, multiply, and spread to the whole of U.S. society—indeed, to the whole world.

Power begets power. In the absence of other political activities, those actors who gain economic power will tend to dominate both markets and the political agenda. Governments will be most concerned with meeting the demands of economically powerful constituents.[13] Environmental protection must, therefore, have its own set of advocates ready to challenge the power elite. Only when political opposition becomes more organized and mobilized around collective forms of resistance will the environmental and social equity agendas come into serious conflict with the dominant group's agenda.

Political resistance requires much more than just "feeling good" about one's efforts. When only a few people join in opposing the powerful, actions that run counter to the dominant group's agenda will encounter enormous pressure to conform to the status quo. Most likely, in acquiescence to this pressure, resistance will stop.[14] When "everyone is doing it," however, the resistance becomes a new dominant mode. This level of environmentalist behavior has not yet been realized in any society.

How and when one might participate in some form of collective action effort to enhance environmental protection is a much more open-ended question. Part 3 of *Environmental Injustices, Political Struggles* provides some insight. Clearly, action requires effort and commitment. Additionally, action depends on one's personality, his or her resources and values, and the social, political, and economic contexts in which the individual is embedded at any given moment. In short, many factors influence what forms participation might take. What is fundamental to taking action, however, is one's inner strength or conviction to do something about a real or perceived problem. At some point, the individual must take a position on what is "right" and what is "wrong." This will require strong moral action. In the context of environmental protection, one can speak of an environmental ethics that leads to collective action efforts by others of a similar predisposition.

Environmental Ethics

Let it be clear: the need for an environmental ethic was the principal point of contention in the last section. The discussion above exposed the need for critical thinking about and understanding of the policies of the private and public sectors that have been developed in response to environmental issues. It challenged the assumptions—for example, the cultural values leading to a "consumer society"—on which those policies are based; it examined traditional solutions and advocated new ways of thinking about the environment; and it appraised the responsibilities toward the environment, and how these responsibilities ought to be reflected in the policies adopted by governments and private companies, as well as in the habits of populations as a whole.

According to some,[15] the traditional approach of Western societies toward understanding ethics may not be suitable within the context of environmental ethics. Anthropocentric definitions of ethics command the attention of Western thought. Ethicists' chief concern is with human behavior that they view as thoughtful, conscious, and intelligent. They assume the individual's ability to discern between right and wrong, good and evil, moral and immoral. In this discernment, the individual chooses a course of action most suited for attaining the "good life." Because the individual is a member of a social grouping, his or her behavior is not purely atomistic in quality. Ethicists, therefore, also pay attention to the actions and practices that aim to improve the individual's life, as well as the welfare of the entire community. Their objective is to clarify what constitutes human welfare, or the "public good," and the kind of conduct necessary to promote such welfare. Concepts and language that are used to direct such conduct are explored. Ethicists then reach conclusions about what goals people ought to pursue and what actions they ought to perform.

Clearly, the traditional approach to the study of ethics is anthropocentric: it says nothing about the welfare of nature and the rights of nature or animals that may interfere with the human quest for the "good life." The result of limiting ethical concerns to human beings is that the environment is subject to exploitation and abuse in the interests of promoting human welfare and meeting the public good. Moreover, the traditional approach to ethics serves the interests of political-economic elites and reinforces the ethical concerns that are fundamental to a materialistic approach to human welfare. The environment is there to be exploited and used for human purposes; there is little concern for environmental sustainability.[16]

As to what is to be done, one could take the opinion that nature must be seen from an ecological rather than an economic perspective.[17] This view is problematic. When there is a choice to be made between an ethical "ought" and a technical "must"—something business must do to remain viable within the marketplace—it seems obvious which path most businesses will follow. Corporate America is locked into an ongoing system of values and ethics that largely determines the actions that can be taken. The competitive context of the marketplace will make the implementation of environmental ethics difficult. Nor can environmental ethics be legislated through regulation of the marketplace or individual economic consumption. Public policy approaches (regulation) experience the same kinds of ethical dilemmas that overwhelm attempts to deal with the social responsibilities of businesses and consumers. Such difficulties of implementation might leave one with the opinion that all of the normative questions about corporate and individual "ethics" are largely unresolved.

Yet there are some changes taking place. There has been a "greening" of industry.[18] Business is beginning to realize that the economy is dependent on the environment; in fact, protecting the environment actually leads to greater economic prosperity and growth. The empirical evidence substantiates this belief.[19] Paul H. Templet finds that states with lower pollution levels and better environmental policies generally have more jobs, better socioeconomic conditions, and are more attractive to new business.[20] This shift in business behavior, however, should not be overstated, for it is only a slight change at best. More important, public policy decisions grounded in the principles of environmental ethics and justice are still lacking. Governmental responses to social concerns are made within the established framework of the traditional economic view. In other words, public policy responses are shaped to correspond with the dominant economic value system that determines corporate behavior and individual consumption. While business is realizing that protecting the environment serves its self-interest, government is lagging behind.

Herein lies the significance of the appeal made to the reader in the first chapter: *you* must take some sort of political action around environmental issues because there is a price for not caring—the quality of democracy is at stake. It is in a collective, direct experience with politics that the whole of society will be better represented and power more evenly distributed.

Environmental Justice and Individual Responsibility

Cultural, political, economic, and social factors that shape human behavior in society have been examined for their impact on the causes of environmental problems. In the final analysis, changing individual behavior is the only way of responding to environmental problems. This individual behavior must spring from the doctrines and principles of environmental ethics and justice. Anthropocentric definitions of ethics that command the attention of Western thought should not be discarded, otherwise the human condition identified in *Environmental Injustices, Political Struggles* will not be addressed. If we can easily treat, consciously or unconsciously, human beings in a discriminatory, racist, and classist way, the environment is in great danger. Conscious and intelligent behavior that guides the individual to discern between right and wrong should remain a chief concern. The ethical conclusion must be that no individual should have to carry an undue, unfair environmental burden of responsibility because of their race or class. A central question posed by ethicists should inform our frame of reference: What actions and practices improve an individual's life as well as the welfare of the entire community? It is hoped that the answer points to individual initiative and responsibility.

In political matters, a passive citizenry invites the abuse of power. Citizens have not been encouraged to use their own initiative and powers of decision making. Consequently, they carry no burden of responsibility. This damages democracy. Here, I extrapolate from Dr. Martin Luther King Jr.'s *Where Do We Go from Here: Chaos or Community,* where Dr. King maintained that racism has no justification in our age. He viewed racism as socially cruel and called for the direct and immediate abolition of racism. It has been suggested in *Environmental Injustices, Political Struggles* that racism is no longer the problem; self-destructiveness is. The politics of power and domination threaten the whole social, political, and economic order. *We are all in this together.*

Notes

1. Rogene A. Buchholz, *Principles of Environmental Management: The Greening of Business* (Englewood Cliffs, N.J.: Prentice-Hall, 1993), 48.
2. Allan Schnaiberg and Kenneth Alan Gould, *Environment and Society: The Enduring Conflict* (New York: St. Martin's Press, 1994), 20.
3. Sherry Cable and Charles Cable, *Environmental Problems, Grassroots Solutions: The Politics of Grassroots Environmental Conflict* (New York: St. Martin's Press, 1995), 124.

4. Lynn White, "The Historical Roots of Our Ecological Crisis," *Science* 155 (1967): 1203–7.

5. The following discussion draws heavily from Dale Westphal and Fred Westphal, eds., *Planet in Peril: Essays in Environmental Ethics* (Orlando, Fla.: Harcourt, Brace and Company, 1994).

6. See, for example, Buchholz, *Principles of Environmental Management;* and Schnaiberg and Gould, *Environment and Society.*

7. Steven Kelman, "Cost-Benefit Analysis: An Ethical Critique," in Westphal and Westphal, *Planet in Peril,* 137–46.

8. For a theoretical presentation of this view see Mancur Olson, *The Logic of Collective Action: Public Goods and the Theory of Groups* (Cambridge, Mass.: Harvard University Press, 1965); and for an application of this view see Terry M. Moe, *The Organization of Interests: Incentives and the Internal Dynamics of Political Interest Groups* (Chicago: University of Chicago Press, 1980).

9. This section borrows heavily from suggestions offered by Schaniberg and Gould, *Environment and Society,* 119–42.

10. Jeremy Rifkin, *Biosphere Politics: A Cultural Odyssey from the Middle Ages to the New Age* (San Francisco: Harper Collins, 1991).

11. Marjorie Kelly, "Are You Too Rich If Others Are Poor? On Ethically Gained Prosperity and Its Obligations," *Utne Reader* (September/October 1992): 67–70.

12. Schnaiberg and Gould, *Environment and Society,* 43–116.

13. Charles Beard, *An Economic Interpretation of the Constitution* (New York: Macmillan, 1913); or Kevin Phillips, *The Politics of Rich and Poor: Wealth and the American Electorate in the Reagan Aftermath* (New York: Random House, 1989).

14. An excellent study of the "cooptive" nature of dominant political forces is John Gaventa, *Power and Powerlessness: Quiescence and Rebellion in an Appalachian Valley* (Chicago: University of Illinois Press, 1980).

15. Westphal and Westphal, *Planet in Peril.*

16. There is a growing literature on the subject of "sustainability." Recommended is C. A. Bowers, *Educating for an Ecologically Sustainable Culture: Rethinking Moral Education, Creativity, Intelligence, and Other Modern Orthodoxies* (New York: State University of New York Press, 1995).

17. See, for example, Daniel Botkin, *Discordant Harmonies: A New Ecology for the Twenty-first Century* (New York: Oxford University Press, 1990).

18. See, for example, Braden R. Allenby and Deanna J. Richards, eds., *The Greening of Industrial Ecosystems* (Washington, D.C.: National Academy Press, 1994); Buchholz, *Principles of Environmental Management;*

Patricia S. Dillon and Michael S. Baram, *Environmental Strategies for Industry* (Washington, D.C.: Island Press, 1994); Paul Hawken, *The Ecology of Commerce: A Declaration of Sustainability* (New York: Harper Collins, 1993); and Stephan Schmidheiny, *Changing Course: A Global Business Perspective on Development and the Environment* (Cambridge, Mass.: MIT Press, 1992).

19. Adam B. Jaffe, Steven R. Peterson, Paul R. Portney, and Robert N. Stavins, "Environmental Regulation and the Competitiveness of U.S. Manufacturing: What Does the Evidence Tell Us?" *Journal of Economic Literature* 33 (March 1995): 132–63.

20. Paul H. Templet, "The Positive Relationship between Jobs, Environment, and the Economy: An Empirical Analysis and Review," *Spectrum* (spring 1995): 37–49.

CONTRIBUTORS

C. RICHARD BATH is Professor of Political Science, University of Texas at El Paso. He has written on environmental issues along the United States–Mexico border for the last thirty years and is a past president of the Association of Borderlands Scholars.

KATE A. BERRY is Assistant Professor in the Department of Geography at the University of Nevada, Reno. Her research focuses on Native American water rights and policy in the western United States. She is the author of *Red Water, White Law: The McCarran Water Rights Amendment and the Struggle for American Indian Water Rights.* She has also published on the topics of western water policy, water conflict management approaches, and environmental education in tribal colleges.

JOHN G. BRETTING is Assistant Professor of Public Administration at the University of Texas, San Antonio. His teaching and research interests include research methods, public budgeting, intergovernmental administration, and urban political economy. He is coauthor of *Evaluation of the Rocky Mountain Institute's Economic Renewal Program in Alamosa, Colorado,* among other publications. He has a long history of grassroots organizing around environmental issues.

DAVID E. CAMACHO is Associate Professor of Political Science, Northern Arizona University. He is the author of *United States Politics and Democracy* as well as numerous articles. His teaching and research interests include race, power, and politics; public and environmental policy; social movement theory; cultural hegemony; and urban politics. He is involved in grassroots organizations and serves as a consultant in the areas of leadership, self-empowerment, and community-building.

JEANNE NIENABER CLARKE is Professor of Political Science at the University of Arizona, Tucson. She is author or coauthor of three books and several articles on natural resources policy, including *Staking Out the Terrain: Power Differentials among Natural Resource Management Agencies.* She frequently does consulting work for federal and state agencies on environmental policy.

ANDREA K. GERLAK is Assistant Professor of Political Science at Guilford College in Greensboro, North Carolina. Her teaching interests are in U.S. government and politics, public policy, and renewable natural resources. Her areas of research and publications include environmental litigation, the "takings" issue, interagency competition over environmental policy, and environmental racism.

PETER J. LONGO is Associate Professor of Political Science at the University of Nebraska at Kearney. His areas of research and publications include water and environmental policy and constitutional law. He is concerned with the connection between fundamental constitutional rights from a comparative approach and a sound environmental policy. He received a grant from the Canadian government to research environmentalism on the Great Plains and was a visiting scholar at Manitoba University in Winnipeg.

DIANE-MICHELE PRINDEVILLE is Assistant Professor of Government at New Mexico State University. Her areas of teaching and research are ethnic and gender politics, and environmental policy. She received the 1994 Ted Robinson Memorial Award presented by the Southwestern Political Science Association for research in ethnic and race politics.

LINDA ROBYN is Instructor in the Department of Criminal Justice at Northern Arizona University. Her teaching and research interests include Native Americans and the criminal justice system, environmental justice and people of color, and state and corporate crime.

STEPHEN SANDWEISS is Instructor in Political Science at Tacoma Community College, Tacoma, Washington. In addition to environmental justice, his primary research interests include the relationship between law and social reform, the social construction of political protest, and the role of legal discourse and legal tactics in the mobilization of environmental justice activism.

JANET M. TANSKI is Associate Professor of Economics at New Mexico State University, where she teaches courses on economic development, with a particular emphasis on Latin America and Mexico. She has several publications on economic development in Latin America. Her current research analyzes structural changes in Mexico's manufacturing industry since trade liberalization began in the mid-1980s.

MARY M. TIMNEY is Associate Professor of Political Science at the University of Cincinnati. She has been President of the Ohio Environmental Council since 1992. Her research has included studies of the managerial capability of local governments in Ohio to deal with hazardous materials and energy policy developments in the state. Her current research is focused on developing a supranational decision model to improve policymaking for global environmental issues. She has a long history of professional experience related to environmental policy in city and county government and in grassroots organizations.

ROBERTO E. VILLARREAL is Professor of Political Science and Associate Vice President for Academic Affairs at the University of Texas at El Paso. He is cofounder of the Hispanic Leadership Institute and the Paso del Norte Policy Institute. He is author and editor of several books including *Latino Empowerment* and *Latinos and Political Coalitions* and has published numerous articles. His research interests involve Latino politics, political leadership, and higher education and administration. He has received a number of honors and awards, among them Who's Who among Hispanics, Who's Who in the West, and Fellow of the Academic Leadership Institute.

HARVEY L. WHITE is Associate Professor of Public Management and Policy in the Graduate School of Public and International Affairs at the University of Pittsburgh. As a National Kellogg Fellow he studied issues of environmental policy and sustainable development in more than twenty countries. He has lectured and written extensively on these issues and holds leadership positions in several national organizations, including the American Society for Public Administration and the Conference of Minority Public Administrators. He serves on the editorial boards of several scholarly journals, including Editor in Chief of the *Journal of Public Management and Social Policy*.

INDEX

Library of Congress Cataloging-in-Publication Data

Environmental injustices, political struggles : race, class, and the environment / David E. Camacho, editor.

p. cm.

Includes index.

ISBN 0-8223-2225-0 (cloth : alk. paper). —

ISBN 0-8223-2242-0 (pbk. : alk. paper)

1. Environmental justice. 2. Environmental justice—Case studies.

3. Environmental policy. I. Cuesta Camacho, David E. (David Enrique)

GE170.E5763 1998

363.7'03—dc21 98-16749